EcoProduction

Environmental Issues in Logistics and Manufacturing

Series editor

Paulina Golinska, Poznań, Poland

The EcoProduction Series is a forum for presenting emerging environmental issues in Logistics and Manufacturing. Its main objective is a multidisciplinary approach to link the scientific activities in various manufacturing and logistics fields with the sustainability research. It encompasses topical monographs and selected conference proceedings, authored or edited by leading experts as well as by promising young scientists. The Series aims to provide the impulse for new ideas by reporting on the state-of-the-art and motivating for the future development of sustainable manufacturing systems, environmentally conscious operations management and reverse or closed loop logistics.

It aims to bring together academic, industry and government personnel from various countries to present and discuss the challenges for implementation of sustainable policy in the field of production and logistics.

More information about this series at http://www.springer.com/series/10152

Arkadiusz Kawa · Anna Maryniak
Editors

SMART Supply Network

 Springer

Editors
Arkadiusz Kawa
Logistics and Transport Department
Poznań University of Economics
and Business
Poznań, Wielkopolskie
Poland

Anna Maryniak
Logistics and Transport Department
Poznań University of Economics
and Business
Poznań, Wielkopolskie
Poland

ISSN 2193-4614   ISSN 2193-4622   (electronic)
EcoProduction
Environmental Issues in Logistics and Manufacturing
ISBN 978-3-319-91667-5   ISBN 978-3-319-91668-2   (eBook)
https://doi.org/10.1007/978-3-319-91668-2

Library of Congress Control Number: 2018941230

Printed on acid-free paper

This Springer imprint is published by the registered company Springer International Publishing AG
part of Springer Nature
The registered company address is: Gewerbestrasse 11, 6330 Cham, Switzerland

Preface

Dynamic and changing market conditions make it necessary for companies to act in networks to maintain their competitive position. For this reason, they have to adapt their own actions to those of other market players. It requires a SMART attitude, which means that today's supply networks should be: Sustainable, Modern, Adaptive, Robust and Technology-oriented. For example, it concerns making decisions about the extent to which a business model should be green or lean. These decisions are related to logistics, IT, environmental issues and networks (vertical and horizontal) relationships management, especially cooperation between suppliers, customers, competitors, and complementors.

The aim of the book is to describe the approaches, opinions, and ideas focused on new and emerging solutions and technologies that might be successfully applied in the configuration, improvement, and management of supply networks in a highly volatile environment of today's global economy. The use of such solutions and technologies should enable the creation of sustainable supply networks (the one in which balance with regard to social, economic, and pro-environmental issues is maintained) which are modern (i.e., the networks implementing the latest solutions in the management and operational spheres), adaptive (i.e., agile and flexible networks), robust (networks in which the degree of sensitivity for unforeseeable changes is relatively low), and which absorb the latest technological solutions.

Both theoretical and practical approaches in this are presented, especially innovative tools, technologies, methods, instruments in supply network management, and case studies from different sectors and different countries. The book covers both economic, organizational, and technical issues connected to the implementation of new solutions in supply networks. Moreover, it includes main problems and challenges which appear by technologies introduction.

The book provides guidance for supply network management researchers and practitioners with regard to the development of SMART supply network in four key research topics:

1. New Technologies Supply Networks—e-supply network management; blockchain applications and technological solutions in logistics on the example of 3PL.

2. Measurement and Improvement of Supply Networks—drivers, opportunities, and challenges of SMART supply networks; measuring performance of adaptive supply chains and importance and methods of handling of losses in transportation.
3. Green Supply Networks—conceptual framework of intra-firm relationships in green supply chain management; green and lean activities of vertically integrated links as a way of creating smart supply networks and the importance of information flow and knowledge exchange for the creation of green supply chains.
4. E-commerce and Digitalization—mutual influence of traditional trading chains and e-commerce; digital consumer needs in digital supply network creation; value for the customer in the logistics service of e-commerce; and the influence of prosumers on the creation and the process of intelligent products flow.

The aim of the first article included in the first part is to present the concept and the potential of the benefits resulting from possessing the information concerning the after-sales usage of the product by the customer. In order to gain this information, the customer's Internet of Things (IoT) data should be introduced into the supply network management. In the next study, the concept of the blockchain and its current applications in logistics and supply networks are presented. In turn, the purpose of the third chapter is to show the similarities and differences in the logistics services offered by DHL to Polish and Ukrainian customers.

The first article of the second part underlines in what ways technological advancement facilitates business relationships and improves the competitive advantage of supply networks. The authors draw attention to the fact that current knowledge employs Activity-Resource-Actor Model (ARAM) in order to recapitulate the drivers, opportunities, and challenges of smart technologies in a supply network. In the next chapter, the problem of supply chain performance measurement, with reference to the concept of adaptive supply chains, is highlighted. The last study of this part presents the benefits of the application of the Decision Support System (DSS). The authors prove that the system, based on relevant mathematical models and algorithms, makes it possible to reduce the number of multiple deliveries.

The third part contains aspects related to the management of green supply chains. The main goals of the articles included in that part are to show the role of information flow and knowledge exchange in the creation of green supply chains, to determine to what extent supply chains are influenced by green and lean ideas within operational and strategic spheres as well as to analyze the relational factors affecting the performance of green supply chains.

The purpose of the articles included in the fourth part is to study the tendencies of the transition of traditional trade networks into e-commerce, to present reflections on the mutual relationships of tools and technologies that shape the consumer's digital needs, to identify which of them contribute to the changes in the consumers' behavior, as well as to present the role that prosumers play in the process of creation and implementation of intelligent products.

Poznań, Poland   Arkadiusz Kawa
                                                                                      Anna Maryniak

Contents

Part I
New Technologies Supply Networks

E-Supply Network Management—
Unused Potential?

Katarzyna Nowicka

*If I'd asked people what they wanted, they would have asked for
a faster horse.*
—Henry Ford.

Abstract SMART supply network is based on Sustainable, Modern, Adaptive, Robust and Technology-oriented approach to the flows management. Information, product and money flows are those of the special interest of all the efforts undertaken by supply chain managers. However in most of the cases they focus on the implementation of the processes from the perspective on the enterprise. Even the holistic approach to the processes development in the network ends when the product becomes the property of the customer. Supply networks can be built in more valuable manner. It can be caused by technology that enables and allows for tracking and analyzing the way of the actual use of the product by the consumer after the good is bought. The aim of this chapter is to present the concept and the potential of the benefits resulting from possessing the information concerning after-sales usage of the product by the customer. To gain on this information the customer Internet of Things (IoT) data should be introduced into the supply network management. The scientific contribution of the article is based on the thesis that the processes of supply chains or networks do not end at the moment of the sale of the product, but in reality they may have their origin in the information on how the product is actually used after it is sold to the customer. It could be done by SMART e-supply network management based on customers IoT data exploring. The study is based on literature review methodology. The EBSCO and Emerald data bases were used for the research purposes.

Keywords E-supply network · Internet of things · After-sales usage of the product

K. Nowicka (✉)
Warsaw School of Economics, Al. Niepodległości 164, 02-554 Warsaw, Poland
e-mail: knowicka@sgh.waw.pl

© Springer International Publishing AG, part of Springer Nature 2019
A. Kawa and A. Maryniak (eds.), *SMART Supply Network*, EcoProduction,
https://doi.org/10.1007/978-3-319-91668-2_1

1 Introduction to Supply Network Management

Christopher [4] defined supply chain as "the network of organizations that are involved, through upstream and downstream linkages, in the different processes and activities that produce value in the form of products and services delivered to the ultimate consumer". The same author also concludes that whilst the phrase "supply chain management" is now widely used, it could be argued that "demand chain management" would be more appropriate. This is to underline that in fact the chain should be driven by the market (customer and the demand), not by suppliers. At the same time word "chain" should be replaced by "network" as there are multiple suppliers, as well as multiple customers in the whole system [5].

However what is the most important when analyzing competitiveness of the supply chains is the integration that allows taking and implementing strategic decisions in the whole system. Integration becomes a vital factor for effective functioning of supply chains to meet the performance objectives like cost, responsiveness, serviceability and agility. Stevens and Johnson [18, p. 22] described supply chain integration as "the alignment, linkage and coordination of people, processes, information, knowledge, and strategies across the supply chain between all points of contact and influence to facilitate the efficient and effective flows of material, money, information, and knowledge in response to customer needs". According to the Supply Chain Operations Reference (SCOR) framework, the integration of demand and supply management across a supply chain takes place through four broad processes like plan, source, make and delivery, which involves flow of material, information and fund. In the context of SCOR model, Hoole [8] put forward the competitive priorities of a supply chain. An effective supply chain management intends to provide consistent quality in response to the market needs with agility while eliminating waste out of the system to reduce cost and generate supply chain surplus. However, it requires access to information on time and an efficient data management for governing the activities and performances and hence having business insight to explore opportunities and overcome difficulties [8].

To enable smooth processes development and continuous of flows, it seems to be evident that Information Technology (IT) has become instrumental and impact on the formulation of competitive advantage of the supply chain or supply network. Ryssel et al. [14] described IT as a technology which enables to communicate, interpret, exchange and use information in the forms of data, voice, images and videos there must be collaboration among both the virtual value chain and the physical value chain for providing required value to the end customers [15]. However Liker and Choi [11, p. 112] commented that "...sharing a lot of information with everyone ensures that no one will have the right information when it's needed". This means that supply chain requires the availability of timely, accurate, precise and relevant information for effective decision making. Therefore the integration aspect and the leader should be engaged.

Information that is important for supply network management is very broad. It includes customer information, sales information, market and competitor information, product and service level requirement, promotion/brand information, demand forecasting, inventory, capacity utilization, process planning and control information, skill inventory, human information, sourcing/vendor information, networking information, logistics, warehouse planning, pricing and fund flow/working capital information, etc. Therefore, the role of data and its management can never be over emphasized in the context of supply network management. Its flows in this system are shown on Fig. 1 [3]. It depicts a typical data driven supply chain structure. Here, demand is initiated by customer end which flows through subsequent stages to the supplier end. Supply of goods and services follows the path from the supplier end to the customer end. Return of goods for repairing, remanufacturing and recycling follows the reverse path to demand [3].

Whole the system is powered by the different types of data and information and would not be feasible to conduct without IT support. Therefore concepts such as e-supply chain, Internet of Things (IoT), smart factory, and industrial internet, have been introduced to represent larger and more complex business systems: from isolated Radio-Frequency Identification (RFID) application to local IoT implementation, to smart factory, and then to part of the global supply chain network within the same company [22]. However, this is a customer who should 'decide' how to build an adequate supply chain or network business model.

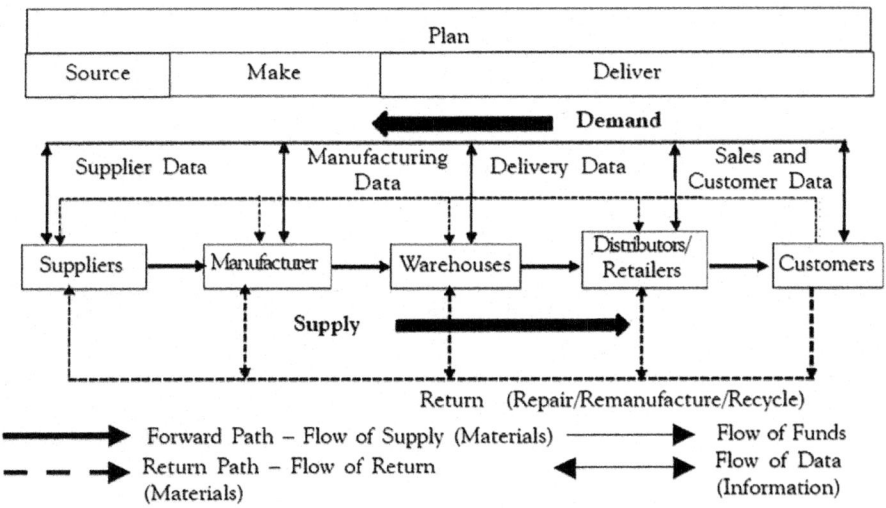

Fig. 1 Data-driven general supply chain structure. *Source* Biswas and Sen [3]

2 Customer Experience and Personalization—the Driving Forces for the Supply Network

Currently instead of designing supply chains from the "factory outwards" the challenge is to design them from the "customer backwards". This means (and is presented on Fig. 1) that the customer is not at the end of the supply chain but at its start [5, p. 39]. As it was already mentioned it is more properly to say demand chain management or demand network management. According to Baker "managing demand chains is fundamentally different to managing supply chains. It requires turning the supply chain on its head, and taking the end user of the organization's point of departure and its final destination." [2]. Therefore an appropriate sequence of actions to create market-driven supply chain might begin with an understanding of the value that customers seek in the market in which the company competes. This sequence is shown on Fig. 2 [5, p. 40].

Customer can be defined in multiple ways, i.e. as end user; the one who is affecting the buying decision; the one who is interacting with the product; the one who is buying, etc. The recognition and differentiation of the customer type, the consumer and the user enables companies in proposing adequate value and matching resources during the cooperation. Those aspects should be revised prior to the decision on how to build the best supply chain (network) business model that meets the needs of right customer or user.

Nowadays, supply chains' managers use different methods for quickly responding to the fluctuations in the demand and customers' needs and expectations. There might be the following concepts introduced into the management of to help gaining on competitive supply chains:

– demand-driven supply chains,
– lean management,
– *leagile* (lean and agile) supply chains,
– involving customers in product design and development, crowdsourcing,

Fig. 2 Linking customer value to supply chain strategy. *Source* Christopher [5]

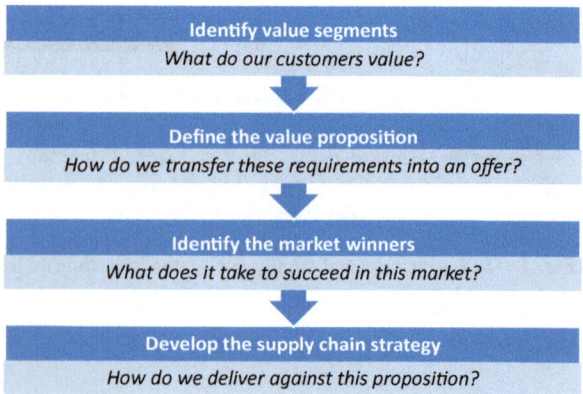

Identify value segments
What do our customers value?

Define the value proposition
How do we transfer these requirements into an offer?

Identify the market winners
What does it take to succeed in this market?

Develop the supply chain strategy
How do we deliver against this proposition?

- analyzing customers journeys and customers experience during the cooperation with the company (mostly supplier),
- customers segmentation,
- introduction customers relation management systems,
- analyzing customers' satisfaction level,
- market researches, etc.

All those activities and solutions are very important. However they might be not sufficient for maximize profits. For example "after mapping five customer segments, one industrial Original Equipment Manufacturer (OEM) found that nearly 70% of its marketing dollars and sales efforts across them were not directed at what mattered most to customers. The company had invested heavily in customized demonstrations to roll out next-generation equipment. The demos were available to all customers, but only those in two of the segments—product enthusiasts and R&D innovators—really cared about participating in them. The rest, comprising over half of the customer base, were happy to visit a plant only occasionally, receive information remotely, or wait their turn for a technical specialist to visit with a standard demo kit" [12]. This actually might mean lack of proper tool for customers' needs identification, their segmentation and marketing.

One of the solution helping companies to impact on their supply chain competitiveness is to reimagine and digitize entire "customer journeys". These are the beginning-to-end processes that customers experience in getting the product (goods or/and services) they need, across whichever channels they choose during the cooperation with the company [6].

Customer experience can be defined as customers' perceptions—both conscious and subconscious—of their relationship with brand resulting from all their interactions with brand during the customer life cycle (www.sas.com/en_us/insights/marketing/customer-experience-management.html). Gartner defined customer experience management as: "the practice of designing and reacting to customer interactions to meet or exceed customer expectations and, thus, increase customer satisfaction, loyalty and advocacy." (www.gartner.com/it-glossary/customer-experience-management-cem/).

To reach this goal companies are creating special programs for supporting customers with the best experience during the 'journey' with the company until the product is sold. To achieve this aim, they are trying to recognize all the needs, wants and expectations of the customers by asking them for that or inviting to design the product at the early stage of its development. The next step is creating personalized experience by integrating advanced digital technologies and proprietary data for customers. Brand individualization unlocks the ability to enhance loyalty with customers by tailoring the experience to each contextual user journey. According to BCG study two-third of respondents representing 50 companies from 10 industries said that they expect at least a 6% incremental annual revenue lift from personalization. Half of the surveyed respondents had more than 25 employees dedicated to personalization programs and were spending more than $5 million a year on personalization campaigns [1]. At the same time companies

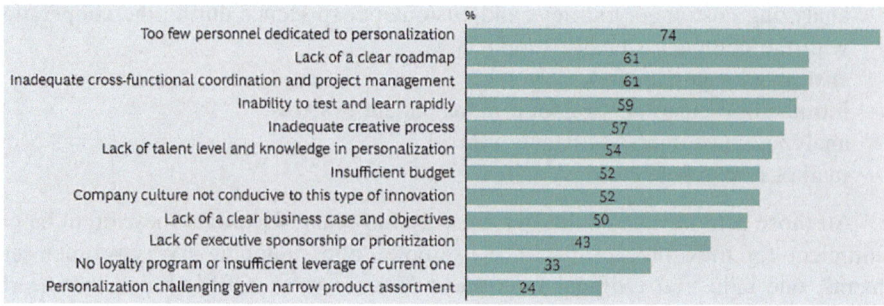

Fig. 3 The top organizational barriers to personalization. *Source* Abraham et al. [1]

indicated a lack of dedicated personnel as the biggest organizational barrier to personalization. Majority of them also faced problems with lack of clear roadmap for personalization and inadequate cross-functional coordination and project management. The main barriers for implementation of personalization in the companies are shown on Fig. 3 [1, p. 4].

To gain on personalization it is important to adopt this concept on the strategic level of the management. It is due to the fact that personalization should not only be a domain of marketing activities, but impacts on the whole business model. It is because it might offer portfolio of different customers' journeys and therefore requires flexible elements of such a business model. The next step is to build data and analytics capabilities in the company or rather in the whole supply chain. To profit on technology the whole system should be transformed to digital solutions based on cloud platform. The last part of the personalization is to enable new ways of working. According to BCG survey, leading companies share common ways of working. For example they collapse silos, create dedicated cross-functional personalization teams, locate their member together and work fast [1].

3   Technology as the Main Driver for Smart Supply Network

Supply chain management may have limited resources to fully recognize advantages that could be obtained by streamlining and improving the entire process of meeting customer needs. Therefore there is a need to move beyond managing supply through a chain of suppliers, manufacturers, distributors, and retailers. The focus on managing fulfillment in a linear way has led to the development of information systems supply chain applications that have primarily supported sequential information flow (e.g. EDI—Electronic Data Interchange; VMI—Vendor-Managed Inventory), and have managed demand information in Customer Relationship Management (CRM) applications separate from supply chain applications. Moreover, the implementation of these applications has often occurred without process change, assuming that the software will directly support existing physical processes and flows [16].

Currently there is a need to focus on improving value for customers in supply or demand networks that should respond to personalized and changing needs. The technology is evolving to accomplish this task. However, its implementation might be stall if trust mechanisms and metrics are not developed to support this new focus on value network advocacy [16]. Anyway, in order to compete by supply network numerous IT systems have been developed. Their goal is to support planning and execution activities. They are implemented within the organization and together with external partners. Those solutions supporting different aspects of planning or executing activities related to fulfillment are shown on Fig. 4 [16]. Inter-organizational coordination was previously accomplished by some early adopters with the use of EDI systems. These systems were examples of "standardization", which involves the establishment of routines or rules that constrain actions of each unit into paths consistent with others in the relationship [19]. EDI standards had to be agreed upon, adopted, and adhered to in order to coordinate the actions of the individual parties.

Today companies have an access to tremendous and powerful volume of data by using digital technologies in their business models. It is structured, unstructured and uncertain. However the most important aspect is the ability to take the advantage from this data. Therefore Big Data Analytics is needed. It should be supported by

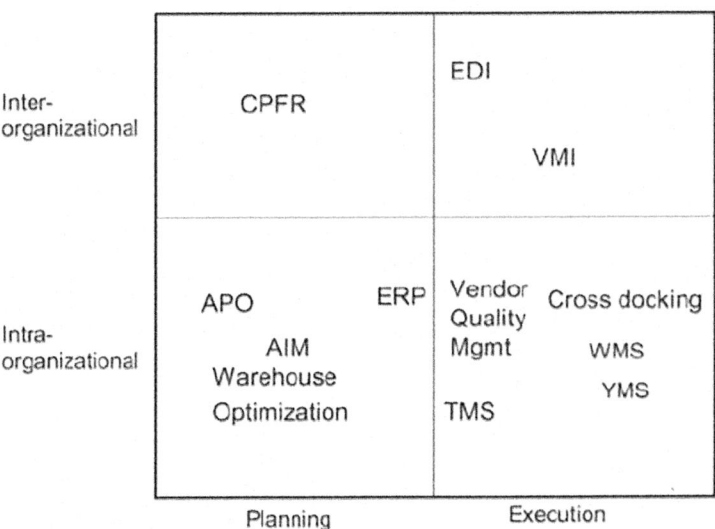

CPFR - Collaborative Planning, Forecasting and Replenishment; EDI - Electronic Data Interchange; VMI - Vendor-Managed Inventory; APO - Advanced Planner & Optimizer; ERP - Enterprise Resource Planning; AIM - Applied & Integrated Manufacturing; WMS – Warehouse Management Systems; TMS – Transport Management Systems, YMS – Yard Management Systems.

Fig. 4 IT investments in the supply chain. *Source* Sherer [16]

the algorithm that enables whole supply network compete with the information impacting on the strategic decisions.

One of the most important sources for data supporting supply chain management is IoT. According to Gartner, IoT is "the network of physical objects that contain embedded technology to communicate and sense or interact with their internal states or the external environment." (www.gartner.com/it-glossary/internet-of-things). Gartner also underlines the IoT Analytics as one of top 10 IoT technologies for 2017 and 2018 saying that IoT business models will exploit the information collected by "things" in many ways. For example, "to understand customer behavior, to deliver services, to improve products, and to identify and intercept business moments. Thus, IoT demands new analytic approaches. New analytic tools and algorithms are needed now, but as data volumes increase through 2021, the needs of the IoT may diverge further from traditional analytics" [21]. At the same time M. Tu recognized four key factors that affect firms' intention to adopt IoT when managing their logistics and supply chain: perceived benefits, perceived cost, trust of technology, and external pressure. According to this study trust of technology indirectly affects the adoption intention through perceived benefits. The research finding indicates that many firms do not feel urgent to adopt IoT when there is no external pressure, such as regulations or strong requirements from customers [20]. Also, for all of the opportunities that IoT offers, there are some significant risks that companies within supply network need to address before they adopt IoT in full measure. There are aspects like regulatory, cybersecurity, privacy, legal, standards, scalability, etc. [7].

Additionally there are several areas that should be fulfilled to gain on data availability through supply networks. First of all those supply networks should have similar features like the ones that characterize smart supply chains. Smart supply chain can be understood as "the new interconnected business system which extends from isolated, local, and single-company applications to supply chain wide systematic smart implementations" [22]. The smart supply chain possess a number of the characteristics, including technologies such as IoT, smart machines, and intelligent infrastructure, and capabilities such as interconnectivity, fully enabling data collection and real-time communication across all supply chain stages, intelligent decision making, and efficient and responsive processes to better serve customers. Smart supply chains collectively possess six distinctive features. They are [22]:

- instrumented: information in is overwhelmingly being machine-generated, i.e. by sensors, RFID tags, meters,
- interconnected: the entire supply chain, including business entities, and assets, IT systems, products, and other smart objects are all connected (i.e. by cloud computing platform),
- intelligent: they are able to make large-scale optimal decisions to optimize performance,
- automated: they must automate much of its process flows by using machines to replace other low-efficiency resources including labor,

- integrated: supply chain process integration involves collaboration across supply chain stages, joint decision making, common systems, and information sharing,
- innovative: innovation is the development of new values through solutions that meet new requirements, inarticulate needs, or even existing needs in better ways.

In a similar way IDC has defined the "thinking supply chain" in the context of five "Cs". Each of these areas contributes critically to the planning of the supply chain as a whole, integrated system [9]:

- connected: being able to access unstructured data from social media, structured data from the IoT, and more traditional data sets available via traditional ERP and B2B integration tools,
- collaborative: IDC has estimated that over 50% of the value creation in manufactured products comes from outside the traditional manufacturing enterprise. Improving collaboration with suppliers is critical, and in the digitally enabled supply chain, this increasingly means the use of cloud-based commerce networks to enable multi-enterprise collaboration and engagement,
- cyber-aware,
- cognitively enabled: the artificial intelligence (AI) platform becomes the modern supply chain's control tower by collating, coordinating, and conducting decisions and next best actions across the chain in an automated and timely way. Certain exceptions would require human intervention, but most of the supply chain would be automated and self-learning,
- comprehensive: analytics capabilities must be scaled with data and in real time. Latency is unnecessary and unacceptable in the supply chain of the future.

The concepts of smart supply chain or thinking supply chain are compatible with the idea of SMART Supply Networks. This is due to the fact that SMART Supply Networks can be characterized as: Sustainable, Modern, Adaptive, Robust and Technology-oriented. This means that they are sustainable supply networks (the ones in which balance with regard to social, economic and pro-environmental issues is maintained) which are modern (i.e. the networks implementing the latest solutions in the management and operational spheres), adaptive (i.e. agile and flexible networks), robust (networks in which the degree of sensitivity for unforeseeable changes is relatively low) and which absorb the latest technological solutions.

All of the described approaches to modern and competitive supply chain (network) have two, basic points of references to gain on value. First—they are strongly supported by digital technology (therefore they might be called "e-supply network") and second—they are customer centric.

4 The Potential of Customer IoT Data for E-Supply Network Management

Today competitive supply networks are the ones that are intimately connected to data sources such as the IoT enabled with comprehensive and fast analytics, openly collaborative through cloud-based commerce networks, conscious of cyber threats, and cognitively interwoven [9]. Supply networks are expected to be data driven and demand aware, they should have access and ability to analyze disparate data sources in the time frames required and profound implication for B2B processes and their underpinning technology. They are e-supply networks. While technology must ultimately serve the interest of supply networks it is extremely important to understand that these technologies will enable new capabilities or new business models across all industries and all regions. However still the most important aspect is to diagnose and differentiate knowledge about needs, wants and expectations of the customer (or rather consumer). Current level of technological maturity is also enabler for recognizing of the actual way of using the product after it is bought by the customer. The information concerning real usage of the product—frequency of usage by the buyer and by other users, type of activity per user and per frequency, connection with other products during its usage, etc. are great undiscovered data, thanks to which the new solutions and product innovations can be built within competitive supply network.

Smart algorithms and machine learning can be used to deliver advanced analytics for new product development. This all can be based on the information that is send by the product and/or shared by the user. In this way users creativity shown on real use-case basis can be the starting point for modeling new solutions and crating additional value proposition. It can be also a new starting point for supply network processes fulfilment of even new business models implementation.

Currently companies collect and use the information from the customer i.e. by tracking and tracing their behavior on web pages or how they share data on social media. Also IoT is a good source of data and information acquisition for portfolio personalization.

The perception of the role of personalization and ability to reorganize actual business model are very important first steps in the way to gain on e-supply network management. However they still might be insufficient to maximize profits from having the relations with the customer or the product user.

It would be also very valuable for the companies and their supply networks to expand the range of common experiences. In most of the cases those customers' experiences journeys end in the moment when the product is sold and for the loyalty reasons some additional incentives are proposed to attract customer for the next shopping. Consumers have increasingly high expectations of the level of customer support and after-sales service. Therefore companies developing services that impact on the quality of brand recognition and experience supporting customers with transportation, possibilities of long-term returns with no costs or repair services, etc. All those activities are very valuable and might be a reasons for customer's choice of the product.

However there are still not too many companies that are collecting and developing knowledge on user experience after the product is bought by the customer. Due to the technology features, today companies have unprecedented opportunities to obtain information about the real usage of products by anybody who has any experience with it. So they are not declarations but real-life cases showing what are the real reasons for buying the product (good). Probably the less complex the product is, the more applications of usage it may have. Analyzing this kind of information can be made by technology implementation and data analyzing. As known, currently the IT companies are tracking usage of web pages or keyboards or what is a customer's path ways in the real (physical) shop. These kind of data is recorded, processed and used for delivering more convenient product or exposing brands in more visible way to the customer.

However, within the product there can be sensors installed that are able to connect with the company and send information concerning services that can or should be supported by the company to the customer. Additionally using IoT supported by sensors, bar codes, Quick Response Code (QR) codes, Global Positioning System (GPS), RFID offer the potential for recognizing new opportunities for value proposition development in SMART supply network. All these data can be collected and processed by cloud computing usage for building innovative supply networks [13].

Cloud computing adaptation is dedicated for processing the huge amount of data (Big Data) collected by various physical assets and for providing flexible access interfaces. Service-oriented methods are applied to facilitate the sharing of assets and services and to enable the development of flexible and scalable Decision Support Systems (DSSs) [27]. According to Yu et al. [28] the IT technology plays an essential role in improving the efficiency and effectiveness of supply chain management. Future technologies like IoT, Big Data Analytics, and Cloud Computing would be possibly adopted to enhance the E-commerce logistics in terms of system level, operational level, and decision-making level that may be real time and intelligent in the next decade.

Additionally there is the concept of smart asset introduced for remote real-time data collection, and the corresponding software agent model is developed to wrap these diverse assets, realizing the Universal Plug and Play (UPnP) working mode [27]. Figure 5 presents architecture of IoT-based tracking and tracing platform, where typical information flow is shown together with different layers engaged in the system [10]. The proposed solution is based on food supply chain and consists of four main layers. The perception layer refers to the physical assets and corresponding smart devices. All physical assets are attached with smart devices to become smart. Smart gateway connects and manages a set of physical assets nearby, processes caches and exchanges real-time data and events locally and temporally, and provides support for service definition, configuration, and execution. The data layer stores the data collected from the perception layer and the execution data generated from the upper layers. Than the service layer is introduced to provide extensive management services in supply chain and excluding the complexity of managing the underlying hardware and software. The last is the application layer that contains applications built upon the services provided by the service layer. Three applications are designed to cover the three key stages of prepackaged food supply chain, from production, logistics, to consumption [10].

The example of the food supply chain architecture of IoT-based tracking and tracing platform based on cloud computing is a good point of reference to further development of SMART supply network knowledge on how products are used and exploited. However this example also shows the flow from the production to the consumption, than the part and the layer concerning data acquisition are the most important to further knowledge development. It can be done by analyzing customer IoT data—the data that is processed after the product is bought by the customer and started to be used by the user.

Some of the companies have already initiated activities of exploring the potential of using customer IoT data to improve products and value proposition. For example there are connected coffee makes already on the market by Nespresso, Starbucks,

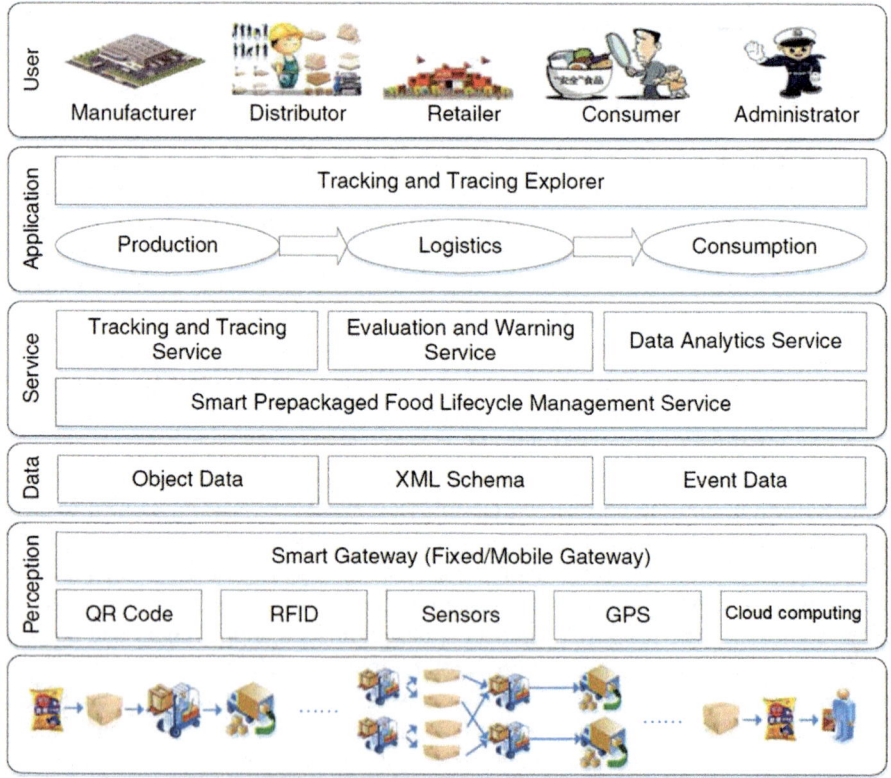

Fig. 5 Architecture of IoT-based tracking and tracing platform. *Source* Li et al. [10]

and others. With customer IoT data coming from the coffee maker, companies can find out i.e. [17]:

- how many cups of coffee customers are making every day,
- whether the coffeemaker is in a warm office or a garage,
- how well it's performing,
- how often it's being used,
- whether the coffee is hot when it comes out of the filter, etc.

One of the sectors that most extensively seeks information concerning customer behavior is the automotive industry. The information how car is driven can be a base not only for designing special features on the next model but also for other companies in supply chain or even in complementary sectors. "The feedback is invaluable for automotive companies. Automakers get to see how customers are using their vehicles. That feedback is the Holy Grail for physical product companies. Before, you had to do surveys from warrantee cards. Then they'd go to maintenance records to see if customers are driving the car hard" [17]. In this way companies can possess information about features of the product that are really important for the customer or user not just theirs opinions or hopes for the product.

For the designer of the product and suppliers of the components, data on how customers use the product can go right into the design of the next generation of the product. It can drive demand network not only based on sales level but on the base of the particular features that are important for the users, so on the base of real-usage behaviors. Such an information has exponential power in gaining supply network competitiveness. It eliminates costs allocated to the activities (and supply network links and intermediaries) that do not add value to the system. It helps offering full personalization based on the most attractive features that are important for particular consumer. Additionally, when connecting different sources of data by IoT usage, e-supply networks are able to get information on the conditions and environment in which product is used. Also the data concerning accompanying or complimentary products that associate usage of the good can be gained. All those aspects opens new potential for innovations in supply network management directly impacting its competitiveness.

Customer IoT data is a new starting point for e-supply network development. The information on how in fact the product is used by the customer might also recreate the customers' experiences and customer journey in a very different way. It can offer him a new solutions that coexists actually not only with the product, but with the each feature of the product that is the most attractive and important for the particular person—customer or user. In this way it reduces supply network risk of offering value proposition that is not in 100% adequate to the expectations and needs of consumer. Therefore the Fig. 2 presented at the beginning of this chapter should have at the beginning the identification of the value based on customer's behavior during the real usage of the product.

Additionally this concept is consistent with SMART supply network idea. The usage of customer IoT data impacts on Sustainability in terms of eliminating wasted materials or other resources that are not important and unnecessary from the perspective of value created for the customer. The concept is Modern in terms of the number of the solutions implemented as practical cases impacting supply network management and business model reconfiguration. This solution also helps supply networks to be closer to the real needs and expectation and when using digital technology it creates Adaptive e-network. This flexibility is a main pillar for Robust supply chains that are able to adapt to any changes actively creating customers' behaviors. It is only possible due to the fact that such a supply networks are based on modern digital solutions and are Technology-oriented.

Of course, as in any other cases, it is not a perfect solution. Especially due to the fact that it is new and not very much popular yet. Therefore, when revising potential of implementation of this concept, the following aspects should be taken into account:

- Data privacy—does the company observe product or individual customer?
- Compliance with actual regulations concerning sharing of data and its protection.
- Ability (technical and mental) to build transparent e-supply networks that requires partnerships—trust and risk sharing. Standards and protocols need to be in the place for the format of data emanating from sensors.
- Ability (technical and mental) to share data between companies from different sectors and different supply networks—building interoperable system that easily connects new links and reconfigure business model.
- Understanding of the role of innovation and ability to quick response to the market or customer needs fluctuations.
- Users willingness to share the data of product usage.
- Supply network should be equipped in the technology that enables smooth data flows (sensors, IoT, cloud computing platform, GPS, QR codes, Big Data Analytics, etc.).

At this level of development, this solution is probably also determined by the supply network that is integrated and has strong, visible leader who is able to implement a new strategy and manage change across the entire network. Therefore this solution might be in special interests of innovators that are not afraid of use unused potential of e-supply network management driven by customer IoT data.

5 Conclusions

When the product or service idea is identified, for the company it is extremely important to recognize the real value that the customer experience when using the product after purchase. The following aspects might be crucial to be able to build business model that supports customers' expectations in the most suitable manner (www.predictableprofits.com/the-10-most-powerful-questions-to-ask-when-developing-a-new-product-or-service/):

- How they would use the product?
- What do they expect from it?
- What problem will this help them to solve?
- Would they be willing to purchase it?
- What value would they put on it?
- What other products in the market place would they consider instead of this?
- What do they like about the other competing products on the market?
- What are the existing products on the market lacking?
- What would help them decide to change to your product?
- Would they even consider using your product if they are already using a competitors?

E-supply network based on the SMART concept might be the solution that helps to recognize personalized needs and adapt resources to build supply system that meets the expectations and eliminates unnecessary costs. This supply network is technology-oriented. Its base for gaining on value is the ability to recognize real type of product usage by the customer after it is sold. To be able to build such a supply network the customer IoT data solution should be introduced.

Customer IoT data is a starting point for modeling e-supply network in the way that plan, source, make and deliver processes are initiated when the particular and important for the company need of the customer is recognized. Those e-supply networks are than driven by real-life usage of the product and not on the opinions, expectations or hopes that customer might have. In this way managers are able to build personalized supply chains that include only justified operations and costs. They can eliminate wasting time, money and therefore develop in a sustainable manner.

In each case this concept must be revised in terms of compliance with regulations and, especially data privacy. It also requires integration within supply chain and a leader that is able to introduce adequate strategy into the whole system.

So, does e-supply network management use its current potential? Today—definitely not. But due to the described reasons, the concept will be probably primarily used by innovators. And it will happen very soon. However it should not be rejected by any of the supply chain manager who wants to crate competitive e-supply chains of the future.

References

1. Abraham, M., Mitchelmore, S., Collins, S., Maness, J., Kistulinec, M., Khodabandeh, S., et al. (2017). *Profiting from personalization*. The Boston Consulting Group
2. Baker, S. (2003). *New customer marketing*. London: Wiley.

 3. Biswas, S., & Sen, J. (2016). A proposed framework of next generation supply chain management using big data analytics. In *Proceedings of National Conference on Emerging Trends in Business and Management: Issues and Challenges, Kolkata, India*. http://ssrn.com/abstract=2755828. January 10, 2018
 4. Christopher, M. (2011). *Logistics and supply chain management* (2nd ed., pp. 1–15). London: Pearson Education.
 5. Christopher, M. (2016). *Logistics and supply chain management* (5th ed., p. 39). London: Pearson Education.
 6. Desmet, D., Markovitch, S., & Paquette, Ch. (2015). Speed and scale: Unlocking digital value in customer journeys, McKinsey, www.mckinsey.com/businessunctions/operations/our-insights/speed-and-scale-unlocking-digital-value-in-customer-journeys#0. January 10, 2018.
 7. EY. (2016). *Internet of Things. Human-machine interactions that unlock possibilities*. Media & Entertainment, EY Global Media & Entertainment Center.
 8. Hoole, R. (2005). Five ways to simplify your supply chain. *Supply Chain Management: An International Journal, 10*(1), 3–6.
 9. IDC. (2017). *The path to a thinking supply chain, technology spotlight*, June:1, 4
10. Li, Z., Liu, G., Liu, L., Lai, X., & Xu, G. (2017). IoT-based tracking and tracing platform for prepackaged food supply chain. *Industrial Management & Data Systems, 117*(9), 1906–1916.
11. Liker, J. K., & Choi, T. Y. (2004). Building deep supplier relationships. *Harvard Business Review, 82*(12), 104–113.
12. Lingqvist, O., Plotkin, C.L., Stanley, J. (2015). Do you really understand how your business customers buy?, McKinsey Quarterly, February
13. Nowicka, K. (2016). Cloud computing a innowacje procesowe w łańcuchu dostaw, in: Zarządzanie łańcuchem dostaw i logistyką w XXI wieku, ed. M. Cichosz, K. Nowicka, A. Pluta-Zaremba, OW SGH, Warszawa
14. Ryssel, R., Ritter, T., & Gemünden, H. G. (2004). The Impact of Information Technology Deployment on Trust, Commitment and value creation in business relationships. *Journal of Business & Industrial Marketing, 19*(3), 197–207.
15. Salo, J., & Karjaluoto, H. (2006). IT—Enabled supply chain management. *Contemporary Management Research, 2*(1), 17–30.
16. Sherer, S. A. (2005). From supply-chain management to value network advocacy: Implications for e-supply chains. *Supply Chain Management: An International Journal, 10* (2), 77–83.
17. Spiegel, R. (2017). Using customer IoT data to improve products, IoT, Design Hardware & Software, October 30. www.designnews.com/iot/using-customer-iot-data-improve-products/66587706557712. January 10, 2018
18. Stevens, G. C., & Johnson, M. (2016). Integrating the supply chain … 25 years on. *International Journal of Physical Distribution & Logistics Management, 46*(1), 19–42.
19. Thompson, J. D. (1967). *Organizations in action: Social science bases of administrative theory*. New York, NY: McGraw-Hill.
20. Tu, M. (2018). An exploratory study of Internet of Things (IoT) adoption intention in logistics and supply chain management: A mixed research approach. *The International Journal of Logistics Management, 29*(1), 131–151.
21. van der Meulen, R., &Woods, V. (2016). Gartner identifies the Top 10 internet of things technologies for 2017 and 2018, STAMFORD, Conn., February 23, www.gartner.com/newsroom/id/3221818. January 14, 2018
22. Wu, L., Yue, X., Jin, A., & Yen, D. C. (2016). Smart supply chain management: A review and implications for future research. *The International Journal of Logistics Management, 27* (2), 395–417.
23. www.gartner.com/it-glossary/customer-experience-management-cem/. January 15, 2018.
24. www.gartner.com/it-glossary/internet-of-things. January 15, 2018.
25. www.predictableprofits.com/the-10-most-powerful-questions-to-ask-when-developing-a-new-product-or-service/. January 10, 2018.

26. www.sas.com/en_us/insights/marketing/customer-experience-management.html. January 10, 2018.
27. Xu, G., Huang, G. Q., & Fang, J. (2015). Cloud asset for urban flood control. *Advanced Engineering Informatics, 29*(3), 355–365.
28. Yu, Y., Wang, X., Zhong, R. Y., & Huang, G. Q. (2017). E-commerce logistics in supply chain management: Implementations and future perspective in furniture industry. *Industrial Management & Data Systems, 117*(10), 2263–2286.

Blockchain Applications in Supply Chain

Davor Dujak and Domagoj Sajter

Abstract Blockchain is a technological concept which evolves from the first cryptocurrency, Bitcoin, and disrupts constantly enlarging areas of economy. The concept of blockchain is developing, and while the future of Bitcoin remains unclear (as it is for the most elements of the economy) it is evident that the blockchain holds enormous potential for large-scale improvements. However, being a technology that could decrease significance many of today's large global corporations, institutions and power structures which have keen interest in preserving established hierarchies, its potential could well remain unexploited. This paper aims to introduce and present the concept of blockchain and its current applications in logistics and supply networks. Blockchain technology promises overpowering trust issues and allowing trustless, secure and authenticated system of logistics and supply chain information exchange in supply networks. The new implementations within supply chain are shifting from blockchain to a wider notion of distributed ledger technologies. Paper presents description and rationale behind current and possible future applications of blockchain in logistics and supply chain.

Keywords Blockchain · Cryptocurrency · Distributed ledger · Supply chain
Logistics

D. Dujak (✉)
Department of Marketing, Faculty of Economics in Osijek,
Josip Juraj Strossmayer University of Osijek, Osijek, Croatia
e-mail: ddujak@efos.hr
URL: http://www.efos.unios.hr/ddujak

D. Sajter
Department of Finance and Accounting, Faculty of Economics in Osijek,
Josip Juraj Strossmayer University of Osijek, Osijek, Croatia
e-mail: sajter@efos.hr
URL: http://www.efos.unios.hr/sajter

© Springer International Publishing AG, part of Springer Nature 2019
A. Kawa and A. Maryniak (eds.), *SMART Supply Network*, EcoProduction,
https://doi.org/10.1007/978-3-319-91668-2_2

1 Introduction

The financial crisis of 2008. fuelled determination of a group of activists to develop a stable, decentralized, autonomous and sustainable financial system, one that would not be under the influence of individual "too big to fail" institutions—moreover—one that would not be under the influence of *any* institution whatsoever. The loss of trust in financial intermediaries which privatized profits but socialized losses motivated tech-savvy enthusiasts to employ internet as by now matured innovation, and significantly powerful (yet affordable) home computers in novel ways. Bitcoin as both payment system and fully digital currency was the first cryptocurrency launched in 2009. Two years later first alternative cryptocurrencies emerged, while at the beginning of 2018 there were more than 1300 of them, beside almost 500 tokens.[1]

One of the Bitcoin's main contributions is the technology of blockchain—its underlying architecture. The concept of blockchain is evolving, and while the future of Bitcoin remains unclear (as it is for the most elements of the economy) it is evident that the blockchain holds enormous potential for large-scale improvements of many different areas of economic system. However, being a strongly disruptive technology that could bring down many of today's large global corporations, institutions and power structures which have keen interest in preserving established hierarchies, its potential could well remain unexploited.

Blockchain has found its applications and is under development in logistics and supply chain activities as well. Radio-frequency identification (RFID), telematics, barcode and 2D codes, sensors-enabled technologies, Internet-of-things (IoT) and numerous other technologies are used for tracking products through supply chain. However, until recently their true potential was not fully exploited as the underlying data was available only within an institution—a company, or perhaps exchanged with limited group of trustworthy partners. Typically, there are numerous supply chain members each with their own information systems, but communication between these systems is limited at best. The main barrier was (and still is) the lack of trust in exchanging information. Blockchain technology promises overpowering trust issues and allowing trustless, secure and authenticated system of logistics and supply chain information exchange in supply networks. Based on these features and blockchain development in general, the pace of new implementations within supply chain is accelerating rapidly. Pilot projects are launched worldwide and supply chain industry is expecting changes.

For majority of companies blockchain is still a mystery when it comes to its practical use in logistic and supply chain activities. This paper aims to introduce and present the concept of blockchain and its current applications in supply chain management. By presenting its characteristics, current applications and future

---

[1]Data from https://coinmarketcap.com (accessed 5. 1. 2018). Tokens are digital assets such as vouchers, debt instruments (IOUs), or real-world objects. They are mostly based on the Ethereum blockchain.

trends, the goal is to provide basic material for academics and practitioners when considering its application in supply chain activities. We attempt to answer following research questions: what is blockchain and how does it function? What are the key features of blockchain applicable in the supply chain and in which supply chain areas are currently being applied? What are future possible development directions for blockchain applications in supply chain?

Paper is structured in four chapters. After the introduction, the second chapter presents the current state of the progress in supply networks. Third chapter analyses the features of blockchain as it came from the cryptocurrency universe, while the next one presents its current implementations and advantages in supply chain and logistics. The fifth chapter concludes.

2   Supply Networks

According to Waters [46, str. 7] supply chain "consists of the series of activities and organisations that materials move through on their journey from initial suppliers to final customers". When organisations "actively (and collaboratively) manage activities and relationships in supply chain to maximize customer value and achieve a sustainable competitive advantage", one can talk about supply chain management, which represents "a conscious effort by the supply chain firms to develop and run supply chains in the most effective & efficient ways possible" [24], str. 8; [7], str. 8; Supply Chain Resource Cooperative, North Carolina State University [41]. Most important supply chain activities are new product development, sourcing, production, logistics, demand management, coordination and integration. In that sense, logistics is part (although biggest) of supply chain management (Vitasek and CSCMP [45]).

Although normative logistics and supply chain management delve into the term of "supply chain", positive economics exhibits that all economic structures which provide products and/or services from the origins to the final consumer are shaped as networks, with numerous participants (supply networks members) on each level and multiple links between them. Therefore, the term "supply networks" (sometimes "supply chain networks" or "distribution networks") is more appropriate since it describes more complex spatio-temporal structures which emphasize the number, position, the nature of relationships, activities, business objectives, capacity, information services and technology base of its participants [33]. Figure 1 represents supply network shown from the perspective of manufacturer.

Supply network of a certain supply chain member (in this case a manufacturer) consists of supply side of network (or supplier network) and of demand side of network (or distributive network). Supply side encompasses all entities of the

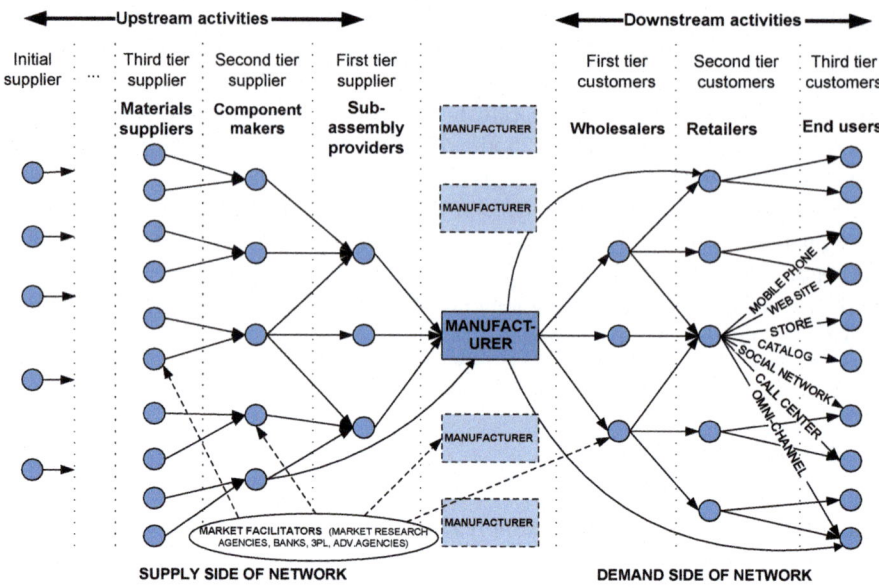

Fig. 1 Supply network around manufacturer. *Source* Revised according to Waters [46]

supply chain that provide inputs, either directly or indirectly to the focal[2] company. The demand side includes all supply chain members that the product passes through on its way to the end consumer [24].

Both supply and demand side of network consist of a certain number of tiers that represent certain supply chain echelons or levels. Conditional on its position in supply network, a focal company can have various tiers on supply and on demand side. Activities carried out on the supply side of the network are referred to as *upstream* activities, while those on the demand side are *downstream* activities. They all focus on improving the flow (primarily material, but also flow of information, services, finance, knowledge, energy, etc.) within the supply network.

Figure 1 also shows the diversity of choices of distribution channels which a retailer (or other member) can use to reach the end user. Moreover, many contemporary retailers simultaneously use multi-channel approach, and some use *omni-channel* retailing approach. Omni-channel retailing is unified and integrated customer-centric experience that allows customers to shop through all possible distribution channels (tablets, smartphones, social networks, kiosks, stores, catalogues, call centres, etc.), at all times [29]. Today there is an increasing number of omni-customers—those that want to be able to buy anytime, any product, in all existing channels (that is, anywhere), through multiple channels at the same time or during the same purchase (e.g. examining the offer on a personal computer, trying

---

[2]Focal company (or company in focus) is supply network member from whose perspective is supply network (or supply chain) mapped.

out at a store, paying over a mobile device to avoid waiting in line at the cash desk, assessing the shopping experience through social networks after the goods were delivered home). In principle, the needs of omni-customers have not changed, but they are now met in new ways, mostly by using information and communication technology (ICT) that also causes changes across the entire supply network. The arrival of omni-customers coincides with the increased need for visibility and traceability in the supply network. Customers in general are constantly pushing new demands to the supply network members, and the customers' demand for feasible, simple insight into the origins of the product and its path through the supply chain has become increasingly common and companies cannot ignore it.

Successful performance in supply networks requires ensuring proper supply network design and continuous optimization of processes that occurs within. Design of the supply network is primarily a strategic, long-term concern. Therefore, when designing a supply network it is necessary to ensure that the supply chain configuration is effective in relation to the expected conditions, but also robust and flexible to adapt to unexpected changes in the surrounding conditions [21]. Design of the distribution or supply network is a form of strategic planning aiming to maximize the economic effects over a longer period of time and also present the consequences of strategic decisions on tactical activities such as the optimization of transport [8, 21]. Facilities in supply network (factories, warehouses, distribution centres, stores) constitute its structure and influence its performance and cost at the same time. While adding facilities enables better customer service (shorter lead time, increased product variety and availability, improved customer shopping experience, increased visibility of supply chain order and increased product return capability), it also means increases in inventory holding and facility costs, and decreases in transportation costs [11]. Therefore, the goal of optimizing the design of the supply or distribution network is to find a trade-off between minimizing the total cost of holding inventory, warehouse costs and transportation costs, while satisfying customer demand related primarily to delivery time. Simply put, network is optimized, "when a minimum of distribution facilities that will meet the customer's response time is reached" [19, str. 188].

There are different approaches to process optimization in supply networks and supply chain management, but this paper focuses on utilization of ICT as the enabler of optimization. Many classifications of ICT in supply chain management exist (see Segetlija et al. 2011). ICTs are often categorized on the basis of its main purpose:

- ICT for data collection and rough processing,
- ICT for data analysis with the aim of making managerial decisions and
- ICT for data integration and exchange.

A challenge of this classification is that new and emerging technologies used in supply chain are now fulfilling more than one purpose at once. In latest report, Gartner [36] gives an overview of emerging technologies in 2017 and most important strategic technology trends for 2018 where such integrative technologies are emphasized; blockchain is emphasized as one of most promising technologies when it comes to logistics and supply chain optimization.

3   Blockchain: Underlying Structure of Cryptocurrencies

The blockchain is a shared, distributed, decentralized, immutable and secure data structure [35]. It is also a protocol for establishing consensus on *valuable information* within a flat network without hierarchy. The "valuable information" can be a transaction with proper authorizations, which is the case for the Bitcoin network, but it can also be some other digital asset, such as property rights. Blockchain can be used as a database, but also as a platform which determines protocols for establishing consensus without a central hub or any intermediary institution.[3]

As a fully digital currency built on the internet for the internet, a Bitcoin is inseparable from its blockchain. They can be regarded as different expressions of the same reality; the blockchain is the consensual database of input/output transactions, and a unit of Bitcoin does not and cannot exist regardless of it. On the other hand, the clever design of the first blockchain is taken from Bitcoin, and its spin-offs are propagated into those areas of economy where there are needs for trust, consensus, security, transparency, storage of value, etc.; one of them being supply networks.

Blockchain relies on encryption. Cryptography can be used to prove knowledge of confidential information without revealing that information, and/or to prove the authenticity of that information. It originates in mathematics, and as such it is employed as a foundation for establishing mutual trust because the clarity, transparency and precision of algebra leave no space for tampering or any interventions. In the cryptocurrency system trustworthiness must no longer be extrapolated from reputational history, tradition, expertise, institutional position or similar sources, but from (complex) mathematics that encrypts information within a virtually impenetrable "shell" of a code. This code can then be easily sent over the internet without fear of anyone making even minuscule modification of the confidential and valuable information content.

A Bitcoin can be defined as "a chain of digital signatures" [34, str. 2]. Within the first cryptocurrency network all communications are entirely transparent and non-encrypted; cryptography is used only for authorization purposes. Every single transaction input must be correctly authorized with a proper digital signature. In this manner a system is created where every communication (with the *money*—Bitcoins —being the substance of communication and sent via transactions) is visible to

---

[3]Intermediaries are not viewed "intruders" per se, but as entities which insert additional layers of costs into the price structure.

anyone, but only the owner of the private key[4] can unlock a transaction output (the address where at some prior point he received the communication—*money*) and use it subsequently as an input in a transaction, thereby redeeming (spending) it.

The crucial element of the arrangement is the algorithm of consensus. Reaching a consensus in a network where by definition a common starting ground is absolute absence of trust among participants, and where each partaker can be a malicious attacker using "insider" knowledge (with the open-source and transparent design), is achieved through controlled, gradual process governed by economic incentives, and through the design of the blocks [4].

The consensus about the rights to redeem information from the blockchain is established step-by-step, where in each step a block with pre-defined elements is constituted. A block contains[5]:

1. reference (pointer) to the previous block,
2. meta-data about the block content which contains the description about the internal structure of the data; a summary of all the transactions in the block[6],
3. the core subject—transaction data, and
4. a random number of specific length called *nonce*.

The four elements above produce together the fifth one: a partly-random value which will serve as a reference in the next block. The reference is partly-random because it is set up to be in a pre-determined interval, but within that interval it can take any value. That reference can be found only by trial and error: by searching through vast variety of nonces until the "winning" one appears. The "winner" produces the link of the current block which will be included into the next one, thereby making a chain.

Therefore, the work that needs to be done consists of:

- the verifications of transactions within a block (checking if the digital signatures are correct and if the user can rightfully redeem a previously received money), and
- finding a winning nonce.

The work is labelled *mining*, and the process is designed so that the miners need to bear hardware and electricity costs in order to be able to find a winning nonce. Their incentive is not grounded within their ideology, altruism, etc., but in new Bitcoins which are disbursed to the winner. Every time a block is found the miner gets rewarded with newly "minted" Bitcoins for finishing first, and the new race for the next block starts. Hence, the system is set up in a manner of a competitive

---

[4]Private keys are an unintelligible string of numbers and letters, thus not revealing the identity of the signatory to other parties in the network, which is another feature of the crypto-network.

[5]This is a simplification to an extent. The description of actual content of a bitcoin block can be found at [4].

[6]I.e. Merkle tree, a data structure used for efficiently summarizing and verifying the integrity of large sets of data [4].

market economy; the more a miner invests in his equipment and the energy that drives it, the more likely he is to find a nonce which proves that he actually did the work.[7]

To sum up (and simplify): mining is blockchain maintenance. It should be clear that a Bitcoin is just a piece of data in the blockchain which does not have any material backing, no underlying asset (not even hardware—memory chips, hard drives, processors—nothing) apart from the electricity spent to provide the proof-of-work. On the other hand, neither do fiat currencies possess real asset backing—they operate as currency because the governments declare them to be money using the historical heritage of paper money which was gradually established as valuable, but also through indolence of the society which accepts the paper as value without rigorous questioning. This provides that cryptocurrencies, albeit categorically virtual and non-palpable, in fact have *some* backing, and it could be argued that they have more backing than modern fiat currencies. In a digital world where electricity powers up all the computers and networks the most appropriate backing for a currency of the future is probably electricity.

Another attractive element of blockchain is its distribution. There is no single authoritative blockchain, no master-node (central server), and no one that controls rogue behaviour. The protocol—predetermined set of rules written in the program code—is the "custodian". Every node can have a complete transaction history[8] which it builds upon and shares among others. This dynamic of sharing the same vision, same protocols, same code and blockchain, but at the same time also competing for new currency through mining, is what gives Bitcoin its resiliency and antifragility.

Altogether, the aforementioned features characterize blockchain technology as a tool for building trust in a network where elements have no other means to establish it. Since the trust is at the core of all financial systems, building trust could be boiled down to building value; since value is money the circle is complete.

3.1 Evolution of Blockchain

The code of the first blockchain (Bitcoin) is open-source, meaning that it is free to use and anyone can see what is behind it and improve on it (given the knowledge and creativity). Anyone can use its architectural concepts and try something different with it, expand features, tinker with the specifications and launch a

---

[7]The question is: searching for a nonce is hard work and costly, but is it meaningful? Pushing the computer to the maximum speed to find an absolutely random number, of what use is that? The use is in providing an existential meaning to money in a digital world. Something cannot come from nothing; digital money cannot be borne out of thin air, from the cloud of zeros and ones. It is "minted" in a computational contest among the competitors which race to verify the contents of the blocks in the chain and to arrange the blocks in a meaningful manner.

[8]In bitcoin that is approx. 150 GB of data, in January 2018.

completely new currency, or fork the original and evolve. This, naturally, opened a vast new space of cryptocurrencies with ever expanding population of designs and users.

The first blockchain was designed to be a transaction protocol and a payment system, and as such its scripting language is intentionally limited: there are no loops or complex capabilities. Every instruction in the Bitcoin code (scripting language) is executed only once, linearly. This ensures that scripts have limited complexity and predictable execution times which is in computer science dubbed as non-Turing-complete[9] programming language. This also means that Bitcoin cannot execute powerful functions that could have infinite loops, which is a security feature intended to protect Bitcoin from malevolent or negligent participants who could otherwise block the network.

However, having an (fairly small and restricted) amount of free space within its data structure, Bitcoin programmers developed applications that could use this space. Although intentionally limited, Bitcoin protocol is not confined to naive input-output transactions; it can convey instructions regarding the transactions such as locking the cryptocurrency for a custom period, and/or requiring multiple signatures for authorization of spending an amount.

Building upon the ideas of first blockchain the developers of what could be called the next generation of blockchain introduced a network—Ethereum—with wider capacity of features. Ethereum's primary innovation was to expand Bitcoin's set of instructions into a fully-featured, Turing-complete programming language. Ethereum is built as a blockchain network which can run different kinds of decentralized applications ("ÐApps" or "dApps"). Even though a precursor Bitcoin could therefore be regarded as a special case of restricted blockchain application, specialized exclusively for transactions (a payment system). A "decentralized application" is sometimes used as a synonym and/or in parallel to "smart contracts". A decentralized application is a server-less peer-to-peer application which runs on a blockchain such as Ethereum, while a smart contract is an agreement written as an algorithm, directly into a code which can be invoked by a decentralized application.

Another innovation of Ethereum are its "sub-currencies". The cryptocurrency of Ethereum is named *ether*, but there are also dedicated *tokens*—sub-currencies (approx. 500 of them) which are used to power specific applications. ERC20 is the standard which defines the characteristics of all Ethereum tokens.

Interventions into the original design of the blockchain bring new features, but does not make newer blockchains necessarily superior [20]. By adding complexity to the protocol Ethereum presents a wider array of points where network-attackers can try to enter or extract data from the blockchain.

---

[9]A programming language is said to be Turing-complete if it can be used to simulate any Turing machine. Turing machine is able to answer any computable problem given enough time and space. The early computers were dissimilar machines because different machines were built for different programs. Mathematician Alan Turing created "Universal Turing Machine" that could take *any* program and run it. Our personal, home computers are universal machines, and they can be used for any program and solve every problem, given enough time and space (memory).

3.2   Types of Blockchain

From the original and still the largest blockchain the idea spun out into the economy encompassing many different designs. At the present point the concepts of blockchain could be categorized in three groups:

1. permissionless,
2. permissioned and
3. private blockchains [37].

Permissionless blockchains (such as Bitcoin, Ethereum, etc.) are decentralized, institutionless, fully public peer-to-peer networks where any member can join without needing a permission from other members.

Permissioned blockchain is similar to the notion of a federation, where a consortium of members permits entrance to new members only with the approval of the existing members. A node needs permission to become a part of the network, as there are barriers to entry in the exclusive club.

Private blockchain is the one where write and/or read permissions are kept centralized to one organization, and as such are disputed to be categorized as blockchains. Furthermore, if we define blockchain as a distributed, decentralized protocol-structure within a flat network without hierarchy, then even the federated, permissioned blockchains cannot be regarded as blockchains.[10] This explains the introduction of a wider concept—the term "distributed ledger technologies" (DLT) which encompasses broader variations of the original design.

Another element of discussion and disagreements is the possibility for the blockchain participants to conspire. One or more elements of the network can try to record false or counterfeit transactions in the new, upcoming block[11] and try to propagate it through the network, but they would need to have the majority of the network (51%) on their side for this attempt to be implanted in everyone's copy of the blockchain. In a permissioned blockchain with few participants it is relatively easy to form a cohort with 51% of votes. That means that the participants would need to trust each other in order not to associate in a >51% pool, but if they trust each other, then the question is do they need blockchain design in the first place. If the inner circle (within the permissioned blockchain) is built from members who trust each other this erases the blockchain's proposition as a trust building mechanism in a trustless network; when members within the network are confident in the behaviour of their peers (hence disallow outsiders to join) it is a question if they need a blockchain, or some version of "distributed ledger technology" would suffice. A permissioned blockchain could be reduced to simpler solutions such as a

---

[10]A case which affirms this position is the introduction of federated blockchain-based technology which replaces the previous clearing system on the Australian Stock Exchange, which is dubbed as "distributed ledger technology", but not as a *blockchain*.

[11]They cannot easily change previous blocks, especially those deeper in the chain; every intervention would be seen because all the subsequent blocks would not have a valid reference.

shared spreadsheet, saved online (in the cloud), with dedicated password for opening and for modification of the content, and with change log[12] [20].

There are numerous open roads with possibilities for further development of blockchain design, with new ones being unveiled regularly. Along with the blockchain the concepts of **side-chain** and **off-chain** emerged.

Sidechains are blockchains with different features from the original blockchain (main chain, usually Bitcoin), which—as the name suggests—are used side by side, in parallel with blockchain, thereby expanding functions and applications of the mainchain.[13] Sidechains are typically connected to the main chain via a two-way peg. A two-way peg is a method of "transferring" Bitcoins from the mainchain to a sidechain; an amount is first locked on an address on the main chain and then activated on the secondary chain. Sidechains are enabled by the more complex Bitcoin functions such as locking script (which "freezes" an amount for a pre-specified time, or until a condition is met) and multiple signatures script (which requires multiple signatures in order to spend an amount).

Off-chain designs broaden the uses of the blockchain by setting up networks which operate certain functions within the blockchain concept, but localizing certain operations away from the blockchain. The Lightning Network is a Bitcoin off-chain protocol where a collateral is made at the opening of the communication channel, and where mutual transactions are recorded locally (not on the blockchain). Only the final transactions, after settlement, are recorded on the blockchain. In a case of dispute a collateral can be taken to indemnify the damaged party. In this way subjects can theoretically have millions of Bitcoin transactions per second for longer periods, but only the final positions are "reported" on the blockchain.[14] The Raiden Network is Ethereum's version of Bitcoin's Lightning Network; an off-chain solution for performing ERC20-compliant token transfers on the Ethereum.

4   Blockchain in Supply Chain

Use of blockchain technology in supply chains and supply networks is fairly new in scientific considerations as first academic scientific papers with this topic appeared in Web of Science and Scopus databases in 2016 [42], and significant growth occurs during year 2017. Based on results of searches in three widely used scientific databases (Web of Science Core Collection, Scopus and EBSCOhost), as well as in the freely accessible Google Scholar it can be concluded that blockchain occupies

---

[12]Google Sheets and Microsoft Excel with Onedrive offer these for free.

[13]Rootstock is a platform for smart contracts in parallel blockchain with Bitcoin (a Bitcoin sidechain), and in this was an alternative to Ethereum.

[14]This could be used for arbitrage between exchanges which could stabilize the price of Bitcoin.

Table 1 Frequency of scientific papers and articles regarding Blockchain and supply chain

Searched term	Article repository			
	Academic databases			Google Scholar
	Web of science[a] (searched within: topic)	Scopus (searched within: article title, abstract, keywords)	EBSCOhost (searched within: title)[b]	
Blockchain	372	787	4.273	16.100
Blockchain and supply chain	**14**	**36**	**70**	**2.870**

Source Authors' search in Web of Science Core Collection, Scopus, EBSCOhost and Google Scholar databases, conducted on January 15th 2018
[a]includes databases of Web of Science Core Collection
[b]includes following databases: Academic Search Complete, Business Source Complete, Inspec, FSTA—Food Science and Technology Abstracts, SocINDEX with Full Text, CINAHL with Full Text, Regional Business News, MasterFILE Premier, Newspaper Source, Library, Information Science & Technology Abstracts, GreenFILE, GeoRef, GeoRef In Process, PsycINFO, PsycARTICLES, eBook Collection (EBSCOhost), CAB Abstracts, EconLit

considerable attention, while its use in conjunction with supply chain is significantly less investigated (Table 1).

Blockchain as a technology in logistics and supply chain is in its developing phase. Even now, it is mostly in the form of tests and pilot projects in many private and public companies, industrial associations (e.g. The Blockchain in Trucking Alliance), followed by or in collaboration with blockchain labs at almost all prestigious universities in the world (e.g. MIT, Columbia, Duke, Berkeley, Cambridge, Cornell, etc.). On a macro level distributed ledger technologies and their uses in supply chain gained considerable attention. Government of the USA recently established an open-source tool for developing and testing blockchain technology services for government, and for public purposes as well [43]. The China Federation of Logistics and Purchasing, the leading logistics association in China, established its own Blockchain Application Subcommittee with main goal of developing standards for blockchain technology [9]. Similar organization are operating in Russia (Russia Blockchain Consortium), Netherland (Dutch National Blockchain Coalition) and many other developed countries.

There are two main characteristics of blockchain technology that are important for its implementation and meaningful use in logistics and supply chains/supply networks (Fig. 2):

- secure, verified, trustable exchange of information through blockchain in real time that makes them accessible to all members of supply network or to anyone else (depending on the type of blockchain),
- possibility of automatic verification and execution of agreed transactions when certain requirements are met through smart-contracts—applications that are living on blockchain [12].

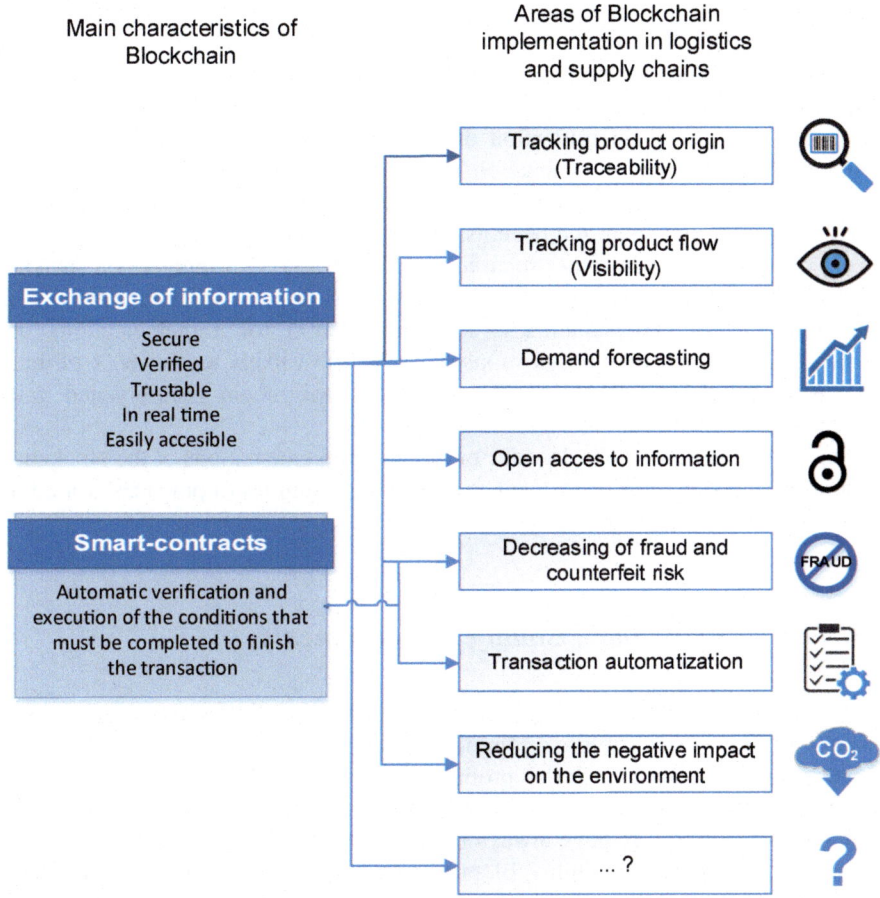

Fig. 2 Characteristics and implementation areas of blockchain in logistics and supply chain. *Source* Authors

Based on these main features of blockchain, implementation areas for its use in logistics and supply chain are being developing in various directions. Some of the most important current implementation areas of blockchain in logistics and supply chain are (Fig. 2) tracking product origin as well as tracking product flow through supply network, demand forecasting, decreasing of counterfeit and fraud risk, open access to information in supply chain, reducing the negative impact on the environment and transaction automatization through smart contracts. In many cases, implementation areas of blockchain are combined in supply chain management, and blockchain is simultaneously used for example for tracking product origin and flow, but also for decreasing fraud risk and more accurate demand forecasting.

A recent study by Hackius and Petersen [23] examined blockchain use in logistics and supply chain management industry on a sample of 152 logistics experts (from consulting, logistics services and sciences; from Germany, United States, Switzerland and France). The results show that companies are still hesitant to dedicate resources for possible blockchain applications. For research purposes blockchain use in supply chain was divided in four areas (use cases): easing paperwork processing, identification of counterfeits and products, facilitation of origin tracking and operationalization the Internet of Things. While in all areas blockchain is evaluated to offer considerable benefits, probability of adoption receives lower ratings than the benefits. Highest benefits are expected from easing paperwork processing. Another focused survey [14] on 42 supply chain management professionals found low level of blockchain awareness between them (this with low level of understanding is also main barrier for implementation), and tracking product moving through supply chain as most likely use for 80% of them. As a main advantage of blockchain improved supply chain visibility and transparency was recognized.

The following paragraphs describe prominent implementations of the blockchain in logistics and supply chain, as well as noticeable examples of practices around the world.

4.1 Traceability and Visibility Enhancements

So far, the most widespread implementation of blockchain for supply chain purposes is in process of verifying origin of product (mostly place, time and who made it) and information about path that products passes from its place of origin all to the way to final consumer (or just from any supplier to any consumer). Questions of traceability and visibility have always been key issues in providing high logistical service to customers. Possibility of providing information where the product is really coming from, who made it, where it was transported, by who and how, or just simply where it is now, is of high value for all customers and true competitive advantage for a company which provide it. On one hand, these information allow better planning and synchronization of customer's processes, leading to further optimizations in operational way. On the other hand, most companies or individuals acting as customers don't really know what is actually happening with products upstream in supply chain, and because of lack of transparency they are actually making less accurate assessment of product's value and are raising questions like: Are these apples really organic? Is this product manufactured truly without children labour? What are true number behind this car's pollution characteristics? Blockchain has ability to provide reliable information to customers concerning product origins and freight route for improved product evaluation before making a decision.

To increase tracking of product origin and way through supply network, blockchain technology is commonly used in pairs with radio frequency identification technology (RFID)—transponders (or tags) on products that carry different product information and are read (or written on) in contactless way through radio waves imitating by different "scanners". The need for use of RFID with blockchain technology (especially in manufacturing supply chain) is highlighted by Abeyratne ands Monfared [1]. RFID system provides the fastest form of non-contact transfer of product information into a digital format—from a product to a computer, or from a computer to a product. It allows you to read information from a large number of products simultaneously, and can record new information on them. The capacity of the information to be written to (depending on the size of the memory chip in the product tag) can be significantly higher than in currently most widely-used product labelling approach—barcode technology. It can include information on origin, place of origin, time stamps, who was responsible for certain activities at generation and distribution, mode of production, places, route and modalities of transportations, temperature of transportation, time constraints of each activity, etc. When RFID started to be used in connection with different sensors on products (e.g. for moisture, movement, temperature, sound, vibration, force, magnetic, acceleration, optical, chemical/gas) and collected information were exchanged through Internet, Internet of Things (IoT) was developed. However, the question of authenticity, verification and security of collected and exchanged information remains insufficiently resolved and regulated. And there comes the blockchain. When the information are collected and transferred to a digital format, blockchain technology enables the verification of this information and entering into a shared distributed ledger, which is complemented and verified in real time. Blockchain provides a form for supply chain mapping [17] and a secure information exchange platform. Data types that are entered in blockchain may be similar to those that are collected by RFID technology, as well as additional data such as environmental impact data, additional processing data or analysis through which the product has passed. The advantage achieved by writing and exchanging with blockchain technology stems from equal visibility of activity and product location information for all members of the supply network and the fact that this information is reliable, secure, authentic and verified only by the members from the supply grid that are authorized to provide them, and no one else can subsequently change it. Other option is to use permissionless blockchain in which case this information become available to any interested party.

Secure product origin tracking complemented by blockchain technology has find its implementation possibilities in many industry, and mostly in food supply chains. Food product and ingredients tracking has special importance when food poisoning, diseases or other forms of contaminations occurs. Blockchain enables much faster and accurate identification of point of origin of problem, followed by recalls and other measures. It is special challenge for food and fast moving consumer good (FMCG) retailers who has to provide all traceability connected information for their customers but almost never have full view of what is happening in upstream part of supply chain. Companies like Walmart, IBM and its partners work on developing standards and solutions for greater safety of food in whole food supply chain by testing tracking and tracing food like pork or mangoes from China to United States. According to Walmart, blockchain supply chain tracking of mango reduced the time it took to trace a package of mangoes from the farm to the store from days or weeks to two seconds [27]. But even if everything functions well, information about food journey can help supply chain members to better prepare for shipment delivery resulting in faster operations and shorter lead time to consumers. Further, consumers have more time to enjoy in product consumption and feel more confident in product that they are eating. Tian [42] has developed an agri-food supply chain traceability system, based on RFID & blockchain technology where RFID serves for implementing data acquisition, circulation and sharing while blockchain technology is used for guaranteeing that shared and published information is reliable and authentic.

Companies from automotive industry (Toyota, Volkswagen, General Motors) are also considering a blockchain technology in different supply chain areas from additive manufacturing to tracking auto parts from the point of manufacturing to assembly plants, but also in self-driving cars sector.

Use of blockchain is not limited just to large companies and their collaboration with ICT giants like IBM. American agricultural conglomerate Cargill is testing blockchain technology to provide to consumers ability to trace the origin of their Thanksgiving turkey back to the farm [40]. Companies like Provenance specializes in developing software solutions powered by blockchain for product traceability and visibility. Provenance solutions enables every physical product to come with a digital 'passport' that proves authenticity and origin [26] *and* that can be tracked through blockchain. Such applications enable to all verified supply chain members to upload and or read information on blockchain as well as final consumers to easily read all this information about product online, in-store or on-pack. Through use of this kind of application, supply chain members insure competitiveness of their products among customers, but also better control of their whole supply chain. Just a simple QR code or RFID tag (on primary packaging of a product) readable by simple app on smartphone can provide access for consumers to all truly relevant data about product origin, processing, transportation, temperature, safety or quality that are inerasably recorded on blockchain. This way consumers have confident in what they're buying and can make true assessment of products quality and value, which ultimately contributes to the growth of loyalty to the brand and the company.

4.2 Improved Demand Forecasting

Demand management is crucial element of supply chain management based on its coordination and integration capabilities. Demand management is not only collaborative approach to planning demand in supply chain, but also includes tools for influencing the demand and supply, by which the demand and supply in the supply chain are adjusted to maximize the profits of the entire supply chain. Demand management in the context of supply chain management can be defined as the preparation of supply chain members for future events in the supply chain through coordinated efforts to forecast expected future demand, jointly influencing demand

and accordingly creating their supply [16]. Most authors agree that demand management consists of following elements:

- demand forecasting and planning,
- supply planning in accordance with demand, and
- collaborative influencing on demand and supply.

Transparency and simultaneously complete security (provided by blockchain) is the basis for successful and long-term information exchange necessary for demand management in the supply network. In a not collaborative managed supply chains there is the risk of losing information in the supply chain—whether they will be provided to competitors or information content will be changed? Therefore each member of supply chain derives their own demand forecast based on available orders from the previous member downstream in supply chain (his buyer) and these demands are called derived demands. This results in a large amount of safety inventory (increasingly rising upstream in the supply chain) which actually are added to the already secured safety inventories on earlier supply chain level. This creates an unnecessarily large amount of inventories in supply chain that financially burdens the chain, slows the material flow and requires unfavourable ways to deal with these excess inventories. This phenomena is called Bullwhip effect and was more popularized by Lee et al. [31].

Supply chain management theory and practice are based on collaborative demand management for avoiding bullwhip effects and optimizing inventory levels through supply chain. In theory all supply chain members (in practice more likely 2 or 3 connected members on different supply chain echelons or levels) produce one common demand forecast based on the data of the independent demand. In supply chains there is only one point of so called independent demand—*the amount of product demanded (by time and location) by the end-use customer of the supply chain* [32]. Either this end-use customer is B2B buyer or final consumer who buys from retailer, he is only one who creates true independent demand for certain products. All other upstream members of supply chain should create their own demand based on this independent demand and that will allow avoiding of multiple additional safety stocks on each new upstream supply chain echelon. Main prerequisite for this common (collaborative) forecast of demand is exchange of data about independent demand between all members of supply chain and main barrier and *crucial problem of contemporary supply networks is lack of trust* for exchanging information between supply chain members. Due to its highest level of system security (from its launch in 2008, Bitcoin's blockchain never "crashed", was never "frozen", nor it was ever hacked) blockchain directly enables solution to fundamental problems for suboptimal achieving of supply network coordination and integration. Savings are possible and can be realized solely on the basis of the belief that the information on original/independent demand is true, available in real time and will not be changed or delivered to competition. Depending on type of blockchain, certain information about independent demand will be accessible only to supply chain members with permission or to any supply chain member (and no one else can delete or change records without other's permission). This way blockchain is becoming *a global system for mediating trust and selective transparency* [10]. Use of blockchain changes a nature of trust. In this trustless network, trust is not connected to a person or a company, but *the burden of trust is within the system ... trust is built in blockchain* [13]. It comes to using the same database without the need for personal confidence—because everyone has the ability to monitor and check the chain for themselves. Buy leaving a problem of trust a side, there is open road for increasing information exchange and trade itself in supply networks. At the same time, even this new trust concept is not taken for granted and therefore trust evaluation models to evaluate enterprises' joint credibility and association credibility under blockchain environment are developing [47].

Additionally, final customer (and/or consumer) could connect to an blockchain-based application, and thus become a "true" member of supply chain with rights and possibilities (finally) to directly express his opinions and needs. Feedbacks from customers could be coming in real time, enabling more accurate forecasting, and radically changing production and retail landscape.

4.3 Open Access

Depending on type of blockchain information/records on blockchain could be available to everyone or just to limited number of participants on distributed ledger. This open access to information in supply chain can provide benefits like ease of paperwork processing, reducing the number of needed direct communications and providing more information to final customer and/or consumer.

When it comes to logistics and supply chain, open access benefits are most recognized in areas of transportation. Maersk and IBM have been developing for some time cargo tracking applications (primarily for containers), as well as application for digitalization of the entire international trade. They started open broad cooperation (with other participants like Microsoft, DuPont, Dow Chemical, Tetra Pak, Port Houston, Rotterdam Port Community System Portbase) from June 2016 enabling container shipping and connected data on blockchain to interested party [28], primarily to insurance companies and banks but as well to all supply chain members, through whole time of goods traveling and by that reducing costs of insurance. At the beginning of January 2018 they announced intention for establishing of global trade digitization platform [44], with tamperproof repository and secure transactions built on open standards of blockchain and designed for use by the entire global shipping ecosystem. Maersk states example when they tested shipping of container of flowers from Kenya to port of Rotterdam requiring around 200 communications between connected organizations during which many waste, spoilage and defects is happening. They also tests international shipments of mandarin oranges from California and pineapples from Columbia. It is estimated that processing of documents and information for container shipments can cost as much as the physical transport itself [44]. By involving all participants of information and material flow into blockchain application and by creating digitized document workflow they managed to ensure all documents and activities in supply chain to be available and visible to every partner, supported with information about who, where and when issued them or move the goods. This decreases the need for domestic and international direct communication, avoids mistakes, waiting, and other forms of waste, and ensure significantly faster information transactions and indirectly faster material flows in supply chain. All information becomes decentralized available reducing delays and various forms of fraud. Main benefit would be accurate and real-time information about the disposition of shipments for ports, terminals, ocean carriers and intermodal transporters allowing them more efficient preparation and planning for their own activities and end-to-end visibility in supply chain activities. According to Marine Transport International estimation blockchain could save $300 per only one container in terms of labour and documents processing [22]. As around 70 million containers are shipped in world every year [3], savings could be considerable. IBM's and Maersk's goal was to start capturing data from 10 million containers by the end of 2017. Additionally, blockchain could help in better optimization of empty containers use through wider access of its availability at nearby ships or ports [15]. Two biggest European port (Rotterdam and Antwerp) have also recognized potentials of blockchain [18]. In future, blockchain will probably extend significantly to other transport modes as well, where it can be paired with some type of existing telematics technology for more secure and transparent exchange of information regarding fleet management.

The use of BC technology in the supply chain can contribute to the more environmentally friendly behavior of both companies and consumers. Most obvious advantage is decreasing of need for paper form documentation enabled through open access and decreased number of online communications and transactions. Additionally, after the lifetime of the product, trustless information about its production and use stages (lifecycle records) can enable more efficient recycling, (re)manufacturing and leasing of existing products [26]. Finally, by tracing carbon footprint of a product using blockchain it is possible to give appreciation to ecologically successful companies and their products, or to penalize the opposite ones. This could be done through charging higher carbon tax or just giving reliable information that will allow consumers to choose not to buy products with higher carbon footprint. Open access to data about products can also significantly increase consumer's trust in it and, as earlier mentioned, can be combined with traceability features of blockchain.

4.4   Fraud Prevention

Verification of authenticity and origin, as well as open access to these data can be strong weapon in combat against fraud and counterfeit products. These blockchain features are especially used in pharmaceutical and luxury jewellery industry.

In pharmaceutical supply chain there are many instances through which medications pass (raw materials suppliers, medical institutions, manufacturers, repackagers, wholesalers, logistics companies, retailers, and patients) and blockchain could help managing such complex supply chain by ensuring medicines visibility and proper reaction in case of need for recalling medicines if problem arises. But still the biggest issue are counterfeit medicines—pharmaceutical market is world's largest fraud market with sales of counterfeit medicines ranging from US $ 163 billion to $217 billion per year according to *PricewaterhouseCoopers (PwC)*—this is especially connected with online purchase of drugs for which the World Health Organization estimates that 50% of the drugs on the Internet are fake [6]. Therefore, pharmaceutical serialisation (prescription drug labelling system for authentication through supply chain from manufacturer to consumer) is practice that becomes mandatory in almost all developed countries—from 2019 will be mandatory in European Union as well (Commission delegated regulation EU 2016/161). Using blockchain as distributed ledger with records on medicines and its origin simplifies serialization and has a potential to significantly decrease this fraud. Consumers could be enabled to choose medicines based on true and verified information from blockchain and to avoid unaware risk for their health arising from use of fake medicines.

Similar combination of blockchain use for traceability and fraud avoiding can be noticed in luxury jewellery industry. Company named Everledger has recognized this need and intents to make the diamond supply chain more transparent, and consequently to reduce fraud, black markets and trafficking. They take 40 metadata points that describe a diamond (e.g. serial number, colour, carats, the cut, the clarity, angles) and they digitally secure records about them on blockchain with linkages to the laser inscription on the girdle of the stone. So far they uploaded 1.6 million diamonds on blockchain platform [39]. Their services are mostly used by insurance companies, banks and open market places in transaction authentication process, and they started to expand their business concept to other luxury goods such as precious wines and artworks.

Increased availability of new technologies usually increases possibility of frauds as well. This is also case with different types of additive manufacturing technologies that allows almost anyone to manufacture individual parts of questionable quality. Kennedy et al. [30] proposed an anti-counterfeiting method coupling lanthanide nanomaterial chemical signatures with blockchain technology for producers and end users to verify authenticity and quality.

4.5 Transaction Automatization

Smart contracts are actually already incorporated in all previously mentioned areas of blockchain use in supply networks and logistics. The main advantages of self-executing transactions on blockchain (smart contracts) are that there are no need for third party (e.g. bank, lawyer or broker) to act as intermediary, and therefore the transaction itself is much faster (especially important for supply chain and logistics) and cheaper, with less possibilities for errors and disruption in execution. Contract in a shape of rules that must be met are embedded in digital code, stored and secured. To create such kind of smart contract, participants in blockchain has to agree about rules as they could be later changed only based on new agreement of them all. Although still mostly in testing phase, in logistics and supply chain management is expected that smart contracts on permissioned blockchain will find its long term place in near future. Bellow will be explained few more complex smart contract applications in logistics and supply chain that are mostly in experimental phase.

In Finland, organization Kouvala Innovation works on experimental approach to connect pallets with shipping tasks, and willing carriers. Pallets are equipped with RFID tags and they communicate their need to be transported to potential carriers on blockchain platform. When best carrier's offer for asked transportation (through mining application) is aligned with requested conditions (price and service), contracts are automatically concluded and executed on blockchain. Carrier is coming for pallets with load and their each move is also recorded [5]. This an example of smart contract on blockchain that carries tasks of smart tendering and smart sourcing.

Delloite [38] argues Blockchains can make supply chain and trade finance documentation more efficient. By providing indisputable level of security for existing digital documents and quick access to all supply chain members, blockchain has a potential to persuade them to leave execution of transactions to smart contracts created on the basis of commonly agreed rules.

Watson IoT Center, Capgemini and IBM work on prototype called Smart Container Management—system that includes containers equipped with sensors sending information on blockchain that are available to all supply chain participants. Different smart contract terms can be activated—e.g. if regulated temperature during transportation decreases below a given threshold, a shipment of replacement products can be triggered in real-time, as well as an insurance proposal, a contractual penalty for the forwarder and a reorder at the supplier [25].

Blockchain in Trucking Alliance association is considering smart contract blockchain application between truckers and brokers that could "automatically provide fuel reimbursements when truckers fill up their tanks, or pay drivers as soon as they deliver their freight" [2], or enable only authenticated drivers to pick up goods (as a way to avoid fake drivers fraud attempt that became frequent in United States).

Table 2 SWOT analysis of blockchain application in supply chain

Strengths	Opportunities
–Decentralization –Transparency –Security –Stability –Automated trust-building system –Automatization of transactions	–Expanding features –New cryptographic functions –New trust-building protocols (proof-of-stake, etc.) –Integration within Internet-of-things
Weaknesses	Threats
–Often not user-friendly –Complex to understand –Many features still in development phase –Achieving consensus regarding system-wide changes within permissionless blockchains –Energy consumption	–Currently mostly unregulated –Lobbying for extensive regulation from intermediary corporations which are being threatened by DLT –Opposition from centuries-old institutions that could be wiped out by DLT

Source Authors

5 Conclusion

Although it is not clear (at this time) if the blockchain is an overemphasized solution[15] looking for the problems it could solve—just another technological innovation which gets people excited but in the end under-delivers—or an actual disruptive force which will sweep across the economy, its potential is certainly unlimited.

The gentle shift in terminology from "blockchain" to "distributed ledger technologies" indicates also distancing and separation from the ideology of the original blockchain designers, notwithstanding that it is the particular worldview that brought the blockchain to mainstream attention. Controlling and taming it could also filter out its main proposition—decentralization.

Blockchain as a technology is not going to replace existing supply chain technologies, but its characteristics of a secure information storage and exchange, as well as automatization of transaction, could assure its place as an important support and upgrade in supply networks (SWOT analysis in Table 2). Regardless of its weaknesses and threats, Blockchain significantly changes information and financial flows that are support to material flows, and thus enables optimization of material flows itself (through cost decreasing and customer satisfaction increasing) as well as raise of exchange based on improved trust in supply chain.

Improving existing and developing new consensus algorithms is at heart of the blockchain future. Cryptography could lay foundation for building trust, as it can

---

[15]Industry is stacked with buzzwords which do have both meaning and purpose, but often get over-used and abused to the point where they become hollow phrases. Examples are *artificial intelligence, big data, cloud-based, disruptive, Internet 3.0* (or higher), *machine learning, internet-of-things, open-source*, etc. Blockchain, unfortunately, seems to be on the same path.

substantially enhance communication between elements in the supply network. Zero-knowledge protocols (a variant of which is a *zero-knowledge succinct non-interactive argument of knowledge—zkSNARK*—already used in some cryptocurrencies) can provide proofs of knowledge of confidential information within certain network elements without revealing this information to the other network participants. Moreover, these protocols can be used to guarantee that communicated data is true and legitimate even though information about the sender, the recipient and other transaction details remain confidential and concealed. Meaningful embedding of these protocols within supply networks could propel them to a substantially higher level.

Bibliography

1. Abeyratne, S. A., & Monfared, R. P. (2016, 9). Blockchain ready manufacturing supply chain using distributed ledger. *International Journal of Research in Engineering and Technology, 5*, 1–10.
2. Aimes, B. (2017). *EDI vendor Kleinschmidt joins blockchain group.* Retrieved December 12, 2017, from DC Velocity: http://www.dcvelocity.com/articles/20170905-edi-vendor-kleinschmidt-joins-blockchain-group/.
3. Allison, I. (2017). *Maersk and IBM want 10 million shipping containers on the global supply blockchain by year-end.* Retrieved December 12, 2017, from International Business Time: http://www.ibtimes.co.uk/maersk-ibm-aim-get-10-million-shipping-containers-onto-global-supply-blockchain-by-year-end-1609778.
4. Antonopoulos, A. M. (2017, 6). *Mastering Bitcoin: Programming the open blockchain* (2nd ed.). O'Reilly Media, Inc.
5. Banker, J. (2017). *Blockchain in the supply chain: Too much hype.* Retrieved December 12, 2017, from Forbes: https://www.forbes.com/sites/stevebanker/2017/09/01/blockchain-in-the-supply-chain-too-much-hype/#6360b7e4198c.
6. Behner, P., Hecht, M.-L., & Wahl, F. (2017). *Fighting counterfeit pharmaceuticals: New defenses for an underestimated—and growing—menace.* Retrieved December 12, 2017, from strategy&.pwc.com: https://www.strategyand.pwc.com/reports/counterfeit-pharmaceuticals.
7. Bozarth, C. B., & Handfield, R. B. (2006). *Introduction to operations and supply chain management.* Pearson—Prentice Hall.
8. Brandimarte, P., & Zotteri, G. (2007, 6). *Introduction to distribution logistics.* Wiley. https://doi.org/10.1002/9780470170052.
9. Buxbaum, P. (2017, 1). Chinese logistics industry enters blockchain era. Retrieved from http://www.globaltrademag.com/global-logistics/chinese-logistics-industry-enters-blockchain-era.
10. Casey, M. J., & Wong, P. (2017, 3). Global supply chains are about to get better, thanks to blockchain. *Harvard Business Review.* Retrieved from https://hbr.org/2017/03/global-supply-chains-are-about-to-get-better-thanks-to-blockchain.
11. Chopra, S., & Meindl, P. (2007). Supply chain management. Strategy, planning & operation. *Das summa summarum des management*, pp. 265–275.
12. Christidis, K., & Devetsikiotis, M. (2016). Blockchains and smart contracts for the internet of things. *IEEE Access, 4*, 2292–2303. https://doi.org/10.1109/access.2016.2566339.
13. Cotrill, K. (2017). *The Benefits of blockchain: Fact or wishful thinking?* Retrieved from chain business insights: Supply chain reimagined: https://www.chainbusinessinsights.com/scm-reimagined.html.

14. DeCovny, S. (2017). *Blockchain in supply chain: Edging toward higher visibility*. Retrieved December 11, 2017, from Chain Business Insights: https://www.chainbusinessinsights.com/blockchain-in-supply-chain-edging-toward-higher-visibility-survey.html.

15. del Castillo, M. (2017). *The world's largest shipping firm now tracks cargo on blockchain*. Retrieved December 12, 2017, from Coindesk: https://www.coindesk.com/worlds-largest-shipping-company-tracking-cargo-blockchain/.

16. Dujak, D., Segetlija, Z., & Mesarić, J. (2017). Efficient demand management in retailing through category management. In P. Golinska-Dawson & A. Kolinski (Eds.), *Efficiency in sustainable supply chains* (pp. 195–216). Berlin: Springer.

17. Duque, A. (2016). *Can blockchains drive supply chain transparency in 2016?* Retrieved December 12, 2017, from LinkedIn: https://www.linkedin.com/pulse/can-blockchains-drive-supply-chain-transparency-2016-andrea-duque.

18. Feuchtwanger, H. (2017). *Logistics on the blockchain? It's happening*. Retrieved December 12, 2017, from Sweetbridge: https://blog.sweetbridge.com/logistics-on-the-blockchain-consider-this-319859d87089.

19. Frazelle, E. (2002). *Supply chain strategy: The logistics of supply chain management*. McGrraw Hill. Retrieved from http://www.academia.edu/download/32893345/book_-_Supply_Chain_Strategy_-_The_Logistics_of_Supply_Chain_Management.pdf.

20. Gerard, D. (2017). *Attack of the 50 foot blockchain: Bitcoin, blockchain, Ethereum & smart contracts*. CreateSpace Independent Publishing Platform.

21. Goetschalckx, M., & Fleischmann, B. (2005). Strategic network planning. In *Supply chain management and advanced planning* (pp. 117–137). Berlin: Springer. https://doi.org/10.1007/3-540-24814-5_7.

22. Grey, E. (2017). *Could blockchain technology revolutionise shipping?* Retrieved December 12, 2017, from Ship Technology: https://www.ship-technology.com/features/featurecould-blockchain-technology-revolutionise-shipping-5920391/.

23. Hackius, N., & Petersen, M. (2017). Blockchain in logistics and supply chain: Trick or treat? *Proceedings of the Hamburg International Conference of Logistics (HICL)-Digitalization in Supply Chain Management and Logistics* (pp. 3–18). Hamburg: epubli.

24. Handfield, R. B., & Nichols, E. L. (2002). *Supply chain redesign: Transforming supply chains into integrated value systems*. Financial Times Press.

25. Heinen, D. (2017). *Blockchain in supply chain management—in the future, trust must be earned rather than paid*. Retrieved December 12, 2017, from Capgemini Consulting: https://www.capgemini.com/consulting/2017/07/blockchain-in-supply-chain-management-in-the-future/#_ftn3.

26. Herzberg, B. (2015). *Blockchain: The solution for transparency in product supply chains*. Retrieved December 12, 2017, from Provenance: https://www.provenance.org/whitepaper.

27. IBM. (2017). *Walmart, JD.com, IBM and Tsinghua University launch a blockchain food safety alliance in China*. Retrieved December 12, 2017, from IBM.COM: https://www-03.ibm.com/press/us/en/pressrelease/53487.wss.

28. IBM. (2018). *Maersk and IBM to form joint venture applying blockchain to improve global trade and digitize supply chains*. Retrieved January 20, 2018, from IBM.COM: http://www-03.ibm.com/press/us/en/pressrelease/53602.wss.

29. Kalakota, R. (2012, 12 1). *Multi-channel to omni-channel retail analytics: A big data use case*. Retrieved 1 20, 2018, from Practical Analytics: http://practicalanalytics.wordpress.com/2012/01/19/omni-channel-retail-analytics-a-big-data-use-case/.

30. Kennedy, Z., Stephenson, D. E., Christ, J., Pope, T. R., Arey, B., Barrett, C. A., et al. (2017). Enhanced anti-counterfeiting measures for additive manufacturing: Coupling lanthanide nanomaterial chemical signatures with blockchain technology. *Journal of Materials Chemistry, 37*(5), 9570–9578.

31. Lee, H. L., Padmanabhan, V., & Whang, S. (1997). Information distortion in a supply chain: The bullwhip effect. *Management Science, 43*(4), 546–558.

32. Mentzer, J. T., Moon, M. A., Estampe, D., & Margolis, G. (2007). Demand management. In *Handbook of global supply chain management* (pp. 65–86). SAGE Publications, Inc. https://doi.org/10.4135/9781412976169.n5.
33. Mesarić, J., & Dujak, D. (2013). Developing supply chain networks—status and trends. *Proceedings of International Scientific Conference Business Logistics in Modern Management* (pp. 59–71). Osijek: Faculty of Economics in Osijek, Josip Juraj Strossmayer University of Osijek.
34. Nakamoto, S. (2008). Bitcoin: A peer-to-peer electronic cash system. Retrieved from https://bitcoin.org/bitcoin.pdf.
35. Narayanan, A., Bonneau, J., Felten, E., Miller, A., & Goldfeder, S. (2016). *Bitcoin and cryptocurrency technologies: A comprehensive introduction*. Princeton University Press. Retrieved from https://lccn.loc.gov/2016014802.
36. Panetta, K. (2017, 10 3). *Gartner Top 10 strategic technology trends for 2018*. Retrieved 1 20, 2018, from Gartner: https://www.gartner.com/smarterwithgartner/gartner-top-10-strategic-technology-trends-for-2018/.
37. Peters, G., & Panayi, E. (2015, 11 18). Understanding modern banking ledgers through blockchain technologies: Future of transaction processing and smart contracts on the internet of money. *SSRN*. doi:https://dx.doi.org/10.2139/ssrn.2692487.
38. Ream, J., Chu, D., & Shatsky, D. (2016). *Upgrading blockchains: Smart contract use cases in industry*. Retrieved December 12, 2017, from Delloite insights: https://www2.deloitte.com/insights/us/en/focus/signals-for-strategists/using-blockchain-for-smart-contracts.html#endnote-21.
39. Roberts, J. J. (2017). *The diamond industry is obsessed with the blockchain*. Retrieved December 14, 2017, from Fortune: http://fortune.com/2017/09/12/diamond-blockchain-everledger/.
48 Segetlija, Z., Mesarić, J., & Dujak, D. (2011). Importance of distribution channels—Marketing Channels—for National Economy, 22nd CROMAR congress: marketing challenges in new economy, (p. 785–809). Pula, Croatia, 6–8 November.
40. Shields, N., & Camhi, J. (2017). *Transportation and logistics briefing: Waymo, GM eye monetizing self-driving technologies—White House announces new commercial drone program—Blockchain's place in agricultural supply chains*. Retrieved December 12, 2017, from Business Insider: http://wwwbusinessinsider.com/transportation-and-logistics-briefing-waymo-gm-eye-monetizing-self-driving-technologies-2017-10.
41. Supply Chain Resource Cooperative, North Carolina State University. (2017, April 02). *Supply chain resource cooperative*. Retrieved February 26, 2018, from What is Supply Chain Management (SCM)? https://scm.ncsu.edu/scm-articles/article/what-is-supply-chain-management.
42. Tian, F. (2016, 6). An agri-food supply chain traceability system for China based on RFID & blockchain technology. *2016 13th International Conference on Service Systems and Service Management ([ICSSSM])*. IEEE. https://doi.org/10.1109/icsssm.2016.7538424.
43. US General Services Administration. (2017, 8 13). *GSA*. Retrieved 1 20, 2018, from Blockchain: https://www.gsa.gov/technology/government-it-initiatives/emerging-citizen-technology/blockchain.
44. Van Kralingen, B. (2018). *IBM, Maersk joint blockchain venture to enhance global trade*. Retrieved January 20, 2018, from IBM.COM: https://www.ibm.com/blogs/think/2018/01/maersk-blockchain/.
45. Vitasek, K., & CSCMP, (2013, August 15). *Supply chain management terms and glossary*. Retrieved February 22, 2018, from council of supply chain management professionals: https://cscmp.org/CSCMP/Educate/SCM_Definitions_and_Glossary_of_Terms/CSCMP/Educate/SCM_Definitions_and_Glossary_of_Terms.aspx?hkey=60879588-f65f-4ab5-8c4b-6878815ef921.

46. Waters, D. (2003). *Logistics: An introduction to supply chain management.* Palgrave Macmillan.
47. Xia, J., & Yongjun, L. (2017). Trust evaluation model for supply chain enterprises under blockchain environment. *Proceedings of the 2017 7th International Conference on Social Network, Communication and Education ({SNCE} 2017)* (pp. 634–638). Atlantis Press. https://doi.org/10.2991/snce-17.2017.129.

New Technological Solutions in Logistics on the Example of Logistics Operators in Poland and Ukraine

Joanna Dyczkowska and Olga Reshetnikova

Abstract The use of modern technologies in logistics was described in this publication. The logistics services of the e-commerce market necessitated an adaptation of the infrastructure to the requirements of customers as regards shipment and collection. The purpose of the study is a presentation of new technologies offered on the territories of Poland and Ukraine as well as their comparison. A literature analysis and a comparative analysis method of the services offered by the largest logistics operator and national mailing services for e-commerce were used in the article. A historical outline and the current development of the e-commerce market in both countries were presented. The sustainable development of the market has resulted in tendencies connected with information technology, ecology and changes in supply chains. The comparative analysis presents the similarity and differences in the logistics services offered by DHL to Polish and Ukrainian customers as well as in the logistics offer of national mailing companies. The direction of changes on the market of logistics services, such as automation of processes, reduction of costs and eco-logistics was indicated in the conclusions, changes to smart supply channel (SSC). The dynamics of the e-commerce market contributed to increased awareness on the part of all the participants of the supply chain and an implementation of changes to improve logistics services provided at each stage of the execution of the order, the results are smart supply network. Nevertheless, there occur differences within the same logistics operator that operates on the Polish and Ukrainian markets. Greater differences in the use of modern technologies can be observed with national mail operators.

J. Dyczkowska (✉)
Faculty of Economic Sciences, Koszalin University of Technology,
Kwiatkowskiego 6E St, Koszalin, Poland
e-mail: jdyczkowska@wp.pl

O. Reshetnikova
Poltava State Agrarian Academy, Poltava, Ukraine
e-mail: olgareshet@ukr.net

47

Keywords E-commerce · Logistics · Logistics service · Operator
E-logistics · Ecologistics · Internet of things (IoT) · Smart supply channel (SSC)
Smart supply network (SSN)

1 Introduction

Professional logistics services on the e-commerce market are most frequently
connected with an adaptation of the infrastructure available to customer requirements. A comprehensive management of the supply chain necessitates the creation
of shipping and receiving points, an advanced storage system as well as appropriate
technological and IT facilities. Logistics operators invest not only in modern
technology but also in the training of their own personnel, which guarantees
long-term cooperation with customers. The e-commerce branch is reporting the
largest growth on the Polish market and yet it still has a huge potential.

The purpose of the study is a presentation of new technologies offered by
logistics operators on the territories of Poland and Ukraine. A comparative analysis
method of the services offered by logistics operators and national mailing companies to e-commerce and a literature analysis were used in this article. The chief
change in e-commerce consists in an acceleration and dissemination of technological progress, which reduces the life cycles of services, such as same-day
delivery, and it forces those companies that sell via the Internet to make changes
concerning the scope of logistics services. The internationalization of the world
economy caused an increase of shipments on the territory of the country and Europe
as well as a high demand for shipments from Asia with a particular consideration of
the Chinese market. There appear new social needs which are reflected in con-
sumerism (e-commerce) and ecology (social responsibility of logistics operators),
as well as the emergence of smart supply networks.

The following hypotheses were put forward:

H1. E-commerce is positively associated with the activities of the logistics
    operator.
H2. Logistics operators implement new technologies in the field of ecology and
    smart supply channel (SSC).
H3. There are differences in the application of modern technologies in logistics and
    national postal operators in Poland and Ukraine.

2 E-Commerce in Poland and Ukraine

Today, it is the Internet that dominates in e-commerce where, obviously enough, it
is not the only one network of data transmission used for this purpose [22, p. 16].
A dynamic increase of the number of e-customers has resulted in logistics operators

becoming interested in this market. Specialized companies emerged on the market that deal with the administration of e-shops and cooperate with carriers [20, p. 37]. The first techniques connected with the exchange of trade documents were based on the mechanisms of Electronic Data Interchange (EDI). These mechanisms enable the transfer of business transaction information between the information systems of companies including invoices, order forms and delivery notes [39]. EDI has been evolving till this day by adapting itself to new networks and data transmission techniques. Initially, electronic commerce would come down to an electronic exchange of data between large companies (e.g. banks, logistics operators) that possessed resources required to build a computer infrastructure to cover their own needs that used EDI. At present, EDI is commonly available and it is used by production, trade and service companies.

E-commerce offers a possibility to pace up business processes, reduce the costs of distribution, to gain new customers and draw up new business models and enter new markets. General agreement in the case of e-commerce is an implementation of electronic markets and digitization of products, where there is an impact on the flow of materials. No clear-cut identification of the scope and directions of effects has been reached as yet, and these questions still remain controversial. Data related to the environmental impact of e-commerce through the use of the Internet is limited. However, the available investigations and examples of e-commerce environmental impact ensure a diversified image of a positive, neutral and negative environmental impact [7]. E-commerce involves a wide array of interactive methods for conducting consumer goods and services. Furthermore, electronic commerce is understood any forms of business operations where parties interact through electronic technologies, and not in the process of a physical exchange or contact [25, p. 256]. M. V. Makarova defines the concept of "e-commerce" as "a kind of business activity in which the interaction of business entities in the sale and purchase of goods and services (both material and informational) is carried out with the help of the global computer network of the Internet or any other information network" [28, p. 272]. A. M. Bereza, I. A. Kozak et al. understand "e-commerce" as the "purchase or sale of goods by electronic means or through a network similar to the Internet". This concept may include orders, payment and delivery of goods or services [6, p. 326].

In 2015, the global population amounted to around 7.3 billion people, of which 1.4 billion people purchased goods and/or services online at least once. In total, they spent $2272.7 bn online, which results in an average spending per e-shopper of $1582. Just like in 2014, Asia-Pacific was the strongest B2C e-commerce region in the world last year. With a B2C e-commerce turn over of $1056.8 bn, it ranked ahead of North America ($664.0 bn) and Europe ($505.1 bn). Latin America and Middle East and North Africa (MENA) were the smallest B2C e-commerce markets in 2015. They achieved B2C e-commerce sales of $33.0 bn and $25.8 bn, respectively. China increased its lead on the United States of America as the country with the highest B2C e-commerce turn over last year. With $766.5 bn, it ranked above the US ($595.1 bn) and the UK ($174.5 bn). Together, these three countries account for 68% of the total global B2C e-commerce turnover. Ranking

Europe in turnover (in million of USD): United Kingdom $174.357, France $72.007, Germany $66.237, Russia $22.785, Spain $20.137 and others $149.604 [9].

The number of Polish Internet users who buy on the Internet has been systematically increasing for several years now. Today, the value of the e-commerce market in Poland is estimated to amount to 36–40 bn zloty ($10.140–11.260 in million), and according to the latest report of "E-commerce in Poland 2017. Gemius for e-Commerce Poland", 54% of Polish Internet users buy online. At present, 26.5 m. of 38.5 m. of the residents of Poland use the Internet. Over a half, that is 54%, have already done shopping online, 53% of the Internet users buy from Polish shops and 16% from shops abroad [44]. Almost half of all the Internet users examined (48%) have declared doing online shopping in the future, 10% of the Internet users buy from foreign e-shops, and the Chinese Aliexpress enjoys the largest success. Looking at the structure of e-shopping users from this group, respondents aged up to 34 with secondary or higher education residing in urban centers and also those who declare good material status of their households constitute the greatest percentage.

Those who do shopping online declare that it is above all convenience, saving time and money and a greater selection of products available than in the case of traditional shops that are the reasons why they use this form of shopping. Those who buy online perceive purchases of this type as little complicated and safe. Having selected products in online shops and websites, the respondents definitely prefer quick transfers via payment services such as Dotpay and PayU as regards payments for products. The users of e-commerce websites, apart from buying on the Internet, very frequently declare searching and comparing the prices of online products via portals such as www.ceneo.pl or www.skapiec.pl. While shopping, they most willingly look for such product categories as clothes (72%), books, records (68%), computer hardware and mobile devices (telephones, smart-phones or tablets) (56%), audio/video devices and household appliances (55%), cinema and theatre tickets (54%), cosmetics and perfumes (51%) as well as shoes (49%). About half of those respondents who do online shopping declare the ROPO effect (research online, purchase offline, which consists in buying products in traditional shops once the shopper has viewed them online). The strength of the effect varies between categories. This effect is the strongest in the case of such products as audio/video devices and household appliances, while it is the weakest in the category of multimedia, computer games and jewelers [40]. Saving of money in the form of lower costs of the purchase of products, special price offers or lower delivery costs of the goods purchased prove to be the most important factor to encourage one to start shopping on the Internet.

Until recently, there was no definition of "electronic commerce" in Ukraine, but Law of Ukraine No. 675-VIII eliminated this gap. Thus, in accordance with this Law, e-commerce is profit-sharing relations that arise in the course of committing transactions relating to the acquisition, change or termination of civil rights and obligations carried out remotely with the use of information and telecommunication systems, resulting in participants in such relations that there are rights and obligations of a property nature [43].

Summary of online purchases of products for Poland and Ukraine (Table 1).

In 2016, almost two thirds of Ukrainian e-consumers (59%), that is less than in the case Polish purchasers, decided to purchase clothes on the Internet. The next positions among the top most willingly purchased products were as follows: household appliances (57%), smart-phones and tablets (57%), transport tickets (54%) and computer hardware (51%) (Table 2).

The Ukrainian e-commerce market is only at the stage of formation, at the same time it has a significant potential for development. According to Kreditprombank (which in turn used materials from Morgan Stanley Research, Fintime, Forbes.ua, Gemius Ukraine, InMind, UIA), the e-commerce market in Ukraine in 2016

Table 1 Comparison of online purchases for Poland and Ukraine

Product	Poland (%)	Ukraine (%)
Clothes	72	58.9
Books, records	68	36.3
Mobile devices	56	56.8
RTV/AGD equipment	55	57.4
Computer hardware	48	50.9
Cinema and theatre tickets	54	47.6
Cosmetics and perfumes	51	35.2
Shoes	49	No date
Sports clothes	46	30.7
Trips, reservation	42	36.6

Source Own statement on based [38]

Table 2 Development of e-commerce in Ukraine, 2007–2016

Indicator/ year	The volume of e-commerce in Ukraine, billion $	Growth (%)	The penetration of e-commerce (%)
2007	0.4	No date	0.6
2008	0.6	50	0.7
2009	0.55	−8	1.0
2010	0.73	34	1.1
2011	1.1	50	1.3
2012	1.59	45	1.6
2013	2.37	49	2.3
2014	3.24	37	2.9
2015	4.44	37	3.8
2016	5.65	27	4.5

Source Sokolenko [41]

amounted to 5.65 billion dollars. However, according to forecasts in 2020, the volume of e-commerce market in dollars will increase more than by twice to reach 14 billion.

As we can see from the table, the growth rates of both penetration of Internet trade into the economy and actual volumes in monetary terms are growing rapidly; however, we are still far away from the level of the TOP-10 countries. It is safe to say that e-commerce has every chance to take up to 80–90% of the share of classical retailing. Moreover, quite a lot will change in other areas: logistics, the market of cashless payment systems, advertising business. Ukraine's trend, which occurred a long time ago in developed economies, is the development of non-cash payments. Now all large platforms work with leading payment systems, which a few years ago was a rare phenomenon. The omnicality in retail is actively developing. To facilitate the development of e-commerce in Ukraine, the network of logistics companies and the development of payment systems and payment methods for buyers can expand. The development of the global Internet, social networks and mobile communications increasingly stimulates the development of e-business. The level of Internet coverage in Ukraine is estimated at 50%, and it is increasing year by year.

3 Ecologistics in Poland and Ukraine

Irresponsible economic activity of people and their exploitative attitude towards nature led to a violation of the ecological balance and gave rise to a number of environmental problems. Logistics as one of the directions of economic activity contributes to the deterioration of the environmental situation, therefore, logistics as a scientific and practical discipline within its activities must take into account environmental aspects in order to minimize the eco-destructive consequences of logistics operations. Ecological logistics is a kind of logistics, the scientific and practical activity of which is aimed at taking into account environmental aspects at all stages of the flow of material and other associated flows in order to optimize resource consumption and minimize destructive environmental impacts [32]. The tendencies of changes on the market of TSL services demonstrate changes in the actions undertaken by logistics operators towards environmental protection. In the formation of the transport and logistic systems of regions or states, an essential role is played by finding a certain optimum between efforts aimed at a reduction of costs in the scale of these systems and efforts aimed at an appropriate level of the service provided and the end customer service. Such an analysis covers an examination not just of the partial elements of the system but rather the provision of comprehensive solutions that offer the possibility of the streamlining of costs. The presented definitions highlight three key characteristics of a SSC (sustainable supply chains). Firstly, more than one entity must be involved in the management of resources, information, and processes that may be beyond a particular company's control. Consequently, the decision-making process includes a number of decision-makers. Secondly, entities partaking in the chain might be working towards contradictory goals, i.e. profit maximisation, carbon footprint reduction, or welfare improvement. The third characteristic aspect is the fact that the environmental impact must be considered in the decision-making process. The carbon footprint of the entire span of the chain must be considered, including suppliers, partners, and clients. Moreover, sustainable development requires adopting an interdisciplinary approach as it necessitates integration of issues and solutions irrespective of functional divisions [24]. The concept of environmental protection in the activities of enterprises connected with transport, shipping and logistics (TSL) is currently acquiring a special significance; this is referred to as eco-logistics or "green logistics". This is testified by an increase of its importance not only in scientific circles but first of all in any work undertaken by logistics operators aimed at environmental protection. Operationalizing green logistics -Green logistics (GL) is measured using a two-item scale based on Murphy and Poist [34] and McKinnon [31], which covers two items: item one refers to choosing the location of the warehouse/distribution center while accounting for emission reduction and renewable energy usage in the center (GL_1). This is related to the second item, that is, using renewable energy efficient lightning, such as sensor lamps and energy-saving lamps, solar power on the roofs, etc.) in the warehouse distribution center (GL_2). In this context, carbon dioxide emissions are frequently cited as a detrimental effect of logistical activities [1, 46]. While the first indicator is more strategic in nature and points to the impact of logistics on the environment, the second is operational and relates to day-to-day operations. Together, they allow a meaningful assessment of green logistics constructs. The results obtained from green supply chain management depend on the level of intellectual capital development of companies. A developed IT system, an effective knowledge diffusion inside and outside of the organization, having certificates supporting supply chain management, a developed motivation system, long-term contracts with clients, a loyalty of suppliers and many more elements contributing to intellectual capital, probably facilitate the green supply management and simultaneously allows for obtaining better results in this area [29]. Therefore, the organizational objectives identified for the implementation of an individual firm's supply chain structure lead to effective conduct that in turn leads to potential achievement of operational and financial goals [5]. Applying the SCP paradigm to sustainable SCM, successful reverse logistics programs have been associated with positive performance measures (e.g. logistics performance [33], economic performance and environmental performance [19].

We follow Zhu et al.'s [48] approach to measure operational performance by assessing the amount of improvement an individual firm achieves on such logistics outcomes as delivery time, inventory levels, and capacity utilization as a result of implementing a sustainable supply chain strategy. Remanufacturing and recycling often provide cost-effective alternatives when compared to the sourcing of new raw materials for use in the supply chain.

Eco-logistics as an integrated system can be described as follows:

- It is based on the concept of the management of the re-circulatory flow of material streams in the economy and the flows of information related to them,

- It guarantees the readiness and ability of an effective planning of segregation and processing as well as recycling of waste according to the accepted process rules and also technical and technological rules that fulfill the standardizing requirements and environmental protection rules,
- It facilitates decision making on technical and organizational levels with the aim of a minimization of the negative effects of the environmental impact that accompany the realization of the processes of supply logistics, re-engineering production, the logistics distributions and servicing in the logistic chains of supplies.

Therefore, an assumption was adopted, that the current situation is connected with a low level of knowledge diffusion within the scope of GSCM [30]. In connection with the above, what seems just is to examine the level of knowledge diffusion within the area of these issues and indicate any troublesome elements connected with it also for logistics operators.

Environmental protection with Deutsche Post DHL—their green solutions are designed to help our customers reach their own environmental targets. Their range of climate-friendly products and services.

Green products and services

1. Carbon Reporting—DHL produce reports on carbon emissions arising from products and services used by the customer, providing an account of the customer's carbon footprint.
2. Carbon Consultancy—DHL analyze their customer's entire supply chain and offer strategies for optimizing transport routes and reducing carbon emissions.
3. Carbon Reduction—DHL offer our customer measures for reducing emissions and saving costs.
4. Carbon Offsetting—DHL offset carbon emissions by investing in officially recognized climate protection projects.

DHL acknowledge that their business has other impacts on the environment such as local air pollutants, their waste production as well as our use of water and paper. Unlike the global impact of carbon emissions, these environmental factors typically impact the environment at the local level. The DHL Group-wide energy efficiency program, together with our ongoing fleet renewal initiative, is helping to minimize our emissions of local air pollutants.

The firm are committed to improving the carbon efficiency of our own operations and those of their transportation subcontractors by 30% compared to our 2007 levels by the year 2020.

It is necessary to note such company as "Nova poshta" in Ukraine. Since 2016, the company has a sustainable development strategy. Over the past year, the company not only set new records in the number of deliveries, but also introduced quality internal changes. Also the company has been the official logistic partner of the eco-action "Let's Make Ukraine Clean!" for several years now and uses its services, resources and the largest network of branches in different cities in Ukraine to help a good cause.

4 New Trends of Logistics Operators in Poland and Ukraine

Direct deliveries to work and home by post or courier services invariably constitute two definitely most willingly selected forms of the collection of products purchased online. The delivery of purchased goods via courier is the form that is most frequently chosen by the respondents: e-buyers. For the activities taken within Internet commerce to be effective and profitable, a lot of attention is to be given to the organization of the logistics process.

E-logistics as one of the forms of e-business focuses on the following issues:

- supply of goods necessary for the proper functioning of the company;
- warehousing and transportation;
- distribution of finished products for certain traders and consumers;
- promotion and offer of a certain range of products, service orders;
- search for new suppliers and clients;
- payment orders.

Advantages of e-logistics:

- it allows to reduce enterprises costs. Thanks to the use of computers and the Internet as well as specialized software, most logistics processes can be implemented at significantly lower costs: related both to trading and communication activities as well as management. It promotes an optimal use of time: it shortens the time of implementation of logistics processes, as well as the preparation of order cycles; it improves the control of the processes;
- it shortens supply chains: this is due to the exclusion of some intermediaries in distribution channels. In addition, it allows one to effectively explore the market and quickly use the results properly;
- information available on the net about business partners helps the company reduce the risk of additional costs that could arise due to an improper choice of business partners.

Logistics in e-commerce can be described as having three aspects: product management including delivery forecasts, storing and delivery of goods to the customer [23, p. 52]. Smooth communication between the customer, the seller and the operator is usually mentioned when talking about those factors that decide about an effective logistics service of online shopping customers. Making information on the order status available to the customer has become a standard in this line of business. The customer may track their shipment either on dedicated websites or by receiving notifications from the operator by e-mail or SMS. When one understands e-logistics as a field of logistics that consists in the use of the Internet and information systems to coordinate and integrate the processes and activities leading to the delivery of products from producers or retailers to consumers, the thesis seems legitimate that online trade has an impact on the logistics industry causing it to change [2, p. 8].

Information about customers is of a key significance to the success of a business model, and it should be used to manage designing of the proposals of values, distribution channels, customer relations and earnings streams [35]. Collecting information about customers should go beyond the current customers to cover potential customers and other market segments, and it should take the external environment into consideration [21]. Understanding who the customer is and what are their values are is of a key significance [37]. For a company to deliver a value to the customer via its network regardless of the fact whether it has a direct contact with the final customer, it must extend its knowledge and understanding of the end customer. The challenge that faces logistics companies consists in gaining a competitive advantage with the aid of information technology to improve customer service and reduce costs at while simultaneously drawing up a new logistics offer. This causes IT to be an important part of the corporate plan of action and one of the grounds of the business potential of a carrier or a logistics operator and their competitive advantage as well as a fundamental part of their business model. Globalization, supply chain management and e-commerce are changing this industry. There is a tendency addressed to producers and distributors who place themselves as strategic partners in outsourcing. Logistics companies used to apply a relatively simple formula based on individual features or combinations of shipments for valuation. Now they had to pay for contracts where it was the service provided that was available and not transport. In many cases, they were asked about "an estimation of the value" by their customers: those prices which could demonstrate not only better performance but also an overall cost saving. In reality, it proved to be a benefit for many of these suppliers as they discovered that after their organization to provide services they can do it at a significantly lower cost than their customers; this not only owing to their knowledge and owing to the joint infrastructure of transport, storage and information systems, but also because their disciplines of quick relocation have allowed the customers to save significant sums as concerns avoiding out-of-date products, quality deterioration and theft as well as a reduction of stock financing costs. There was just a considerably smaller amount of awaiting stocks or stocks transferred between various distribution levels. Therefore, the price of the value was substantially more advantageous to both parties [42]. Although there is little data on the impact on production and provisions, there is some detailed research of the impact of e-commerce on logistics. The main problems connected with new electronic trade business models involve energy and packaging materials used by logistics networks for the realization and delivery of products.

Logistics operators must pay attention to the productivity, which has a direct impact on environmental protection depending on such parameters as shipping distances, freight value for return, allocations of purchases, population density, quantity of packagings and transport type. Simulations show that the deliveries of groceries home may reduce car traffic from 2 to 19%, electricity consumption from 5 to 35% and carbon dioxide emissions from 7 to 90% depending on the context and assumptions [10]. These investigations also emphasize that some direct effects of electronic food shopping contribute to a reduction of road traffic, and it is possible to identify indirect effects in relation to the natural environment. From the ecological perspective, the key question is whether e-commerce should be the first choice and what type of influence on environmental protection may improve the activity of the chain of values through better use of the capacity of vehicles, avoiding express delivery by air or minimization of packagings [3].

Quick and efficient delivery is one of the most important factors that influence the selection of an online shop. The "FlexDeliveryService" service is adapted to market tendencies, and recipients can make changes flexibly in the course of transport of consignments with the use of the latest technologies (the Internet, the smart-phone). It is a system solution owing to which consignees may freely decide about the date, place, method, delivery term of online consignments and information as to whether the company will deliver the consignment to them; this is without any need to contact the helpline, the courier or its branch. Offering flexible delivery of goods contributes to a greater probability of choosing the shop again by online shoppers. The "FlexDeliveryService" service offers many free options of delivery and changes in communication with the consignee. Logistics operators consider the "same day delivery", i.e. the same day delivery to the customer who placed the order [49] to be the greatest challenge connected with the growing Internet commerce market. This requires fundamental adaptations in the current supply chain.

Logistics companies implement the Internet of things (IoT) technology, which was mainly used to enable communication between machines and to improve transport effectiveness. It is expected, however, that IoT will play a greater role in the future of logistics and it will increase speed, it will reduce waste volumes and it will lower overall costs. Many people believe that IoT will enable communication with other new technologies, such as AIDC (automatic identification and data capturing), RFID (radio frequency identification) or Bluetooth to identify those elements which need to be changed in accordance with requirements on the part of companies and customers. Designing of the supply chain and software for modelling is so important. Software for supply chain designing performs compound mapping and analyses based on unique requirements, and it calculates benefits for each scenario. It is predicted that the world's industry of Semiconductor will exceed the threshold of USD 400 bn for the first time in the year 2017. There had been an increase of the market by +16.8% by the year 2016, mainly by developing the market of memory and a quick implementation of those devices that support IoT. DHL experts maintain that the IoT market will slow down the pace of annual income growth by the year 2019. Such a dynamic and unstable market in combination with a high activeness of fusions and takeovers as well as a complexity of electronic supply chain management requires logistics solutions for the whole industry.

5   Smart Supply Chain (SSC)

Logistics operators need to diversify their activities based on new levels of agility, precision and performance. Supply chains need to react to the environment that is becoming increasingly globalize, complex and multi-channeled. Under the pressure of profits, supply chains face the challenge of changes oriented onto customer needs; apart from this, they are oriented onto an increase of online sales and growing demand for prompt deliveries and accuracy. For example, one-hour delivery of products will soon become common in cities not only in Western Europe but in Poland, too. New markets, as well as changes in the models of operators' activities, rely on changes in the processes of relations. Logistics operators perceive smart supply chains as a challenge or an opportunity to convert the supply chain into real competitive advantage. Those logistics companies that do so increase the value of their offers on the market. What is the key to this goal is an integration of various channels, efficient cooperation between participants and real-time smart management, use of the potential of new technologies that have an impact on and optimize integration, predictive intelligence and traceability. Transformation of the traditional supply chain into smart is unavoidable. Promptness, accuracy and efficiency of the service will decide about a competitive advantage. Any company that forms a part of the supply chain in any of its links will be affected by this change. This is in particular true of those logistics operators that provide services on the e-commerce market and companies in the B2B supply chain.

SSC is the approach that manages the movement of raw materials into an organization and it is deployed in both indoor and outdoor environments and the information updates about the goods are uploaded in the server with the help of IoT. IoT refers to the wireless communication between the objects and it can be controlled and monitored from anywhere, any place and at any time. SSC keeps the record of the movement of goods from the supplier to the manufacturer which moves along with the wholesaler to the retailer and finally to the customer. SSC is mainly used to benefit for the company and it links and collects the overall data collected by the warehouse management system, communication software and distribution management system as well. SSC is mainly used in order to increase the efficiency of the company and also to make sure that the materials reach to the customer [47].

In order to meet the challenges facing today's complex supply chains, new concepts for supply chain planning and control are emerging. Typical characteristics of such concepts are [2]:

- Demand-driven control throughout supply chains, basing operations on pull rather than push principles
- Integrated and automated operations
- The use of unified supply chain control models
- Intelligent and advanced information processing, data mining, visualisation and decision-support
- Information sharing and transparent information flow.

The smart supply chain will provide more value to companies and promote new collaboration models amongst actors, with complete end-to-end traceability. All the actors within the chain (from customers to the very last link) will be connected via smart platforms to establish new relationships and share processes and information, leading to optimized costs, delivery times and services (Fig. 1).

(1) Collaborative integration—comprehensive visibility of the chain starting from the purchase of raw materials to the sale; it is possible owing to the integration of information between various links and use of the platform of cooperation. The customer is yet another link in the chain with a complete visibility of all the other stages (e.g. the order status). Fluctuations in the demand affect the management of the entire chain (e.g. automatic initiation of repeated supplementation of orders once products have been removed from the line).

(2) Complete traceability—sensors and communication between all the elements and real time monitoring ensures comprehensive visibility of the SSC. This comprehensive traceability makes it possible to detect gaps and it leads to dynamic and continuous improvement, especially on the stages of transport. Traceability will close the chain and will radically reduce the time required to launch the product on the market, and it will adapt products to customer needs.

(3) Logistics intelligence—by using huge real time information analysis, the organization is learning itself and establishes intelligent systems in the key areas of the chain: network planning, supplies and warehouse management. Complex scenarios and interactions will be dynamically simulated. The SSC model will be optimal all the time, and it will interpret the key parameters both of the environment and customers.

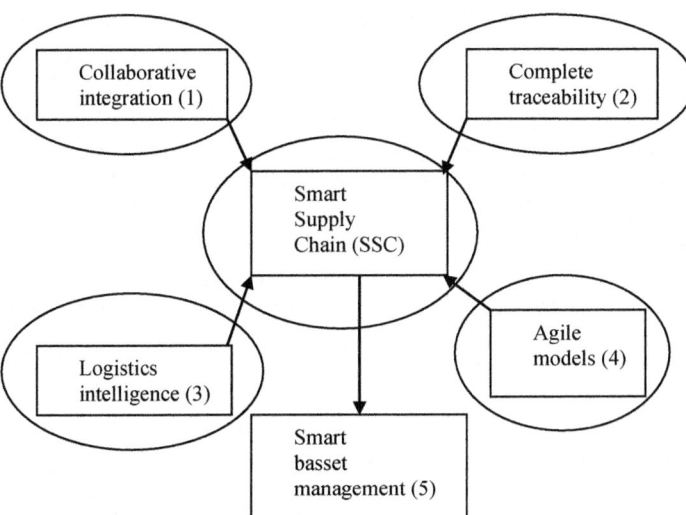

Fig. 1 The condition of the existence of SSN. *Source* Author's own based on [16]

(4) Agile models—predicting of trends in the behaviour and adaptability of an uncertain environment represents an exponential growth of the level of services. Advanced models of demand forecasting are defined, and adaptive models are used to promptly respond to information from the environment and to new demands (e.g. defining of flexible points of delivery: DPD and DHL: "predict and pickup point"). New sale channels and models are addressed to e-commerce customers in particular. Senders gain efficiency owing to continuous cost reduction, greater reliability and minimization of risk to the end customer.

(5) Smart basset management—The assets in your logistics chain are monitored and managed through predictive intelligence algorithms to radically reduce downtimes, thus maximizing use and cutting maintenance costs. The customer will maximize the return on your investments by SSC.

The SSC consists of different streams whereby every stream provides particular information related to the entire contracting process. Underlying the principles of the SSC each stream stores its data on various servers to secure the inviolability of important process data. The extent of the permissions given to the contracting parties is defined by the business community and varies between full access on data, reading-permission only or the complete refusal of all information. This procedure ensures that only a verified party can add data or change status. It also simplifies the traceability in case of data abuse [45]. Apart from the payment cloud, further smart B2B service, such as financing or an insurance service, may be integrated within the confines of an SSC that is managed by a logistics operator, where the entire system is based on a cloud. Based on a comparative analysis in Poland in Ukraine, the customers of logistics operators gain the following [26]:

- those companies that are served by logistics operators operate with greater flexibility while reducing the lead time up to 30% and dynamically adapting the chain to new operating models, new products, new intermediaries and new markets;
- they manage operations more effectively by lowering logistics costs by as much as 20% per annum owing to sustainable planning, reduction of costs, optimization of stocks, simplification of processes and creation of the dynamics of constant improvement;
- within the confines of SSC, logistics companies guarantee greater security and better services by obtaining entirely safe and reliable transport from end to end; by bringing the entire chain to the end customer; by increasing sales by as much as 10% owing to the improved reliability and level of services as well as a flexible response to the demand;
- those companies that are served within the confines of SSC gain greater profitability by rationalizing the portfolio of their products, reducing products with low rotation by as much as 20% and optimizing current assets to maximum.

While logistics operators increasingly speak of the digitization and digitalization of logistics and supply chains with the IoT (enabling multiple logistics use cases and, along with other technologies such as cloud computing and edge computing or fog computing if you prefer, bringing intelligence to the edge across logistics) [25], digital supply chains and (semi-) autonomous decisions and logistical assets such as self-driving trucks and so forth, the role of people in supply chain management is far from over. Digital technologies are providing opportunities to better support the customers of logistics operators with proactive location and delivery information. They also give us better insight into the performance of the smart supply chain. The digital platform merges global transportation data and combines it with relevant weather data and news feeds from traditional sources as well as social media. This picture, which is much more comprehensive, enables our logistics and customer service teams to collaborate on events impacting the smart supply chain, resulting in more timely and accurate information for customers, creating the smart supply network in Europe.

This procedure ensures that only a verified party can add data or change the status. It also simplifies the traceability in the case of data abuse. When a contract component or requirement is fulfilled, the DCU (decentralized control units) transfer the information of completion to one of the related SSC. Subsequently, the smart contract triggers the determined payment. Logistics operators are included in the entire delivery, control and payment procedure. Changing the place of the delivery on the order of the parties.

The next key trend is a more prominent role of the Internet of Things (IoT) in extended smart supply chains. IoT can help logistics operators provide improved predictability of customer demand with real-time visibility of product and service demand signals. In a smart supply chain, strategic deployment of IoT technologies can improve asset utilization, customer service, working capital deployment, waste reduction and sustainability. Real time communication between machines, factories, logistic providers and suppliers provides improved visibility on the end-to-end SSC. IoT can address compliance, regulatory and quality reporting requirements such as parts traceability and product genealogy, emissions and country of origin. With IoT, organizations are better suited to track shipped products for warranties, returns and predictive support for maintenance. The real premise of IoT-enabled SSC is to delegate decision making on some of the operational aspects to smart objects and systems, based on real time analytics and machine learning algorithms. This brings us to a customer- and market-driven smart supply network (Fig. 2).

Organizations would need not only to be aligned with the true customer demand but also to shape the demand using technology and analytical tools. SSN are extremely complex organisms, and no company has yet succeeded in building one that is truly digital. Today, smart supply chains are transforming into dynamic, interconnected systems: smart supply networks by logistics operators. These digital supply networks integrate information from many different sources to drive production and distribution, potentially altering manufacturer's competitive landscape.

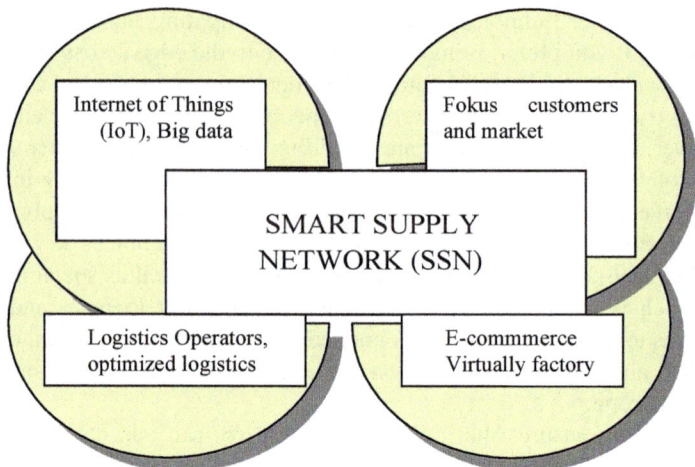

Fig. 2 The technologies and value in connected logistics and the smart supply network. *Source* Author's own study

6 Comparative Analysis

Large shipping companies, such as DHL, UPS, DPD or GLS, extend their services of deliveries and returns to include new functionalities. They also extend the networks of partner points or devices where shipments can be directed to (immediately or after an unsuccessful attempt by the courier to deliver them), or where returns can be made. At the same time, smaller shipping companies are developing; they offer niche services that integrators are not interested in: for example, X-press Couriers has introduced nationwide same day deliveries [36, p. 92]. Those companies that specialize in other segments of logistics or mail markets joined in the battle for e-commerce customers, yet these are those companies that possess experience in logistics industry and suitable resources to handle e-commerce and have a vision of innovative services. Some of these have been introducing new innovative solutions for several years now. For example, InPost has developed a network of parcel stations, Ruch offers collection in retail points of sale; the Polish Post has decided to develop a network of collect and send systems in their post offices. Companies are developing which supply collect and send systems that are installed in shopping centers, petrol stations, in locations close to public transport nodes or office blocks. An example of these includes SwipBoxes manufactured by NoaTech (DHL is one of those that are introducing these).

DHL (13%), DPD (11%) and UPS, Polish Post, In-Post and GLS (3%) are the most preferred brands in Poland by on-line shoppers as regards courier services.

The largest player on this market, which functions on both markets: DHL (Table 3), and national postal operators: the Polish Post and UkrPost (Table 4) were accepted in the analysis.

Table 3 Analysis of services offered by DHL in Poland and Ukraine

Service for e-commerce	DHL in Poland	DHL in Ukraine
Track shipment (track and trace)	Yes norm	Yes norm
Interactive messaging (predict)	Yes 3,25 PLN	No
Find a service point (Pickup)	Yes norm	Yes norm
Personal collection at the facility (click and collect)	Yes norm	Yes norm
Comfort and flexibility of deliveries (last mile)	Yes for an extra charge	No
International shipping (cross-border)	Yes determined individually	Yes determined individually
Market platform service (fulfillment)	Yes for an extra charge	Yes for an extra charge
Omnichanel	Yes for retail	No
eReturn shipment	Yes 13 PLN	Yes
Cash card pickup or Blik	No 8 PLN cash collection by courier	No cash 12,00 UAH collection by courier
Internet sale (netlivery)	Yes	Yes
Common label SSCC (serial shipping container code)	No	No
Supplies management	Yes for retail	Yes for retail

Source Own work based on [12–14]

Professional operators of the logistics market in Ukraine, which mainly serve parcels related to online sales, form a segment that is identified abroad as CEP (courier, expedition, post), even though today it is difficult to clearly assign packages to courier, postal or logistics services. These industries compete with each other, and the differences are increasingly smoothed out. Each of the operators offers their services. For example, courier operators, are expanding the service "from door to door", offer: deliveries on the same day, deliveries for a certain hour; city parcels. Logistic operators, such as DHL, offer deliveries up to a certain hour (until 9:00, until 10:30, until 12:00, until 18:00) and by the end of the working day.

Also, logistic operators guarantee the refund of the tariff in the case of non-compliance with the delivery time, which indicates high standards of customer service. Another example is the Cargo Express company delivering door-to-airport and airport-to-airport services. Additional services include SMS messages about the arrival of the parcel to the destination, delivery of cargoes in non-working hours, at a specified hour, delivered to a specific person, etc. DHL Ukraine offers Time Definitive Delivery to certain specific destinations [4]. Processing several million customs entries is no easy task. DHL Express in Ukraine does that every day for its customers because it acts as a broker on their behalf.

Once thoroughly checked, your shipment information is handed over to the appropriate customs authority. They work with them to expedite the clearance and get your goods delivered quickly and efficiently. In the postal logistics market,

Table 4 Analysis of services offered by Polish post in Poland and Ukrposhta in Ukraine

Service for e-commerce	Polish post	Ukrposhta
Track shipment (track and trace)	Yes norm	Yes norm
Interactive messaging (predict)	Yes 2,60 PLN	No
Find a service point (pickup)	Yes norm	No
Personal collection at the facility (click and collect)	Yes norm	No
Comfort and flexibility of deliveries (last mile)	No	No
International shipping (cross-border)	Yes international price list	Yes international price list
Market platform service (fulfillment)	No	No
Omnichanel	Yes for retail	No
eReturn shipment	Yes 3,70 PLN	Yes 6,00 UAH
Cash card pickup or Blik	No 2 PLN cash collection by courier	No cash 10,00 UAH collection by courier
Internet sale (netlivery)	Yes	Yes
Common label SSCC (serial shipping container code)	No	No
Supplies management	No	No

Source Own work based on [11, 16–18]

Ukrposhta, the largest state logistics company, is still ranked first. According to the Ukrainian Association of Direct Marketing, it owns more than 30% of the market. Over the past 2 years, the company has bought new cars, launched express delivery and made an interface for integration with online stores.

Unlike a standard package, with a delivery period of three to six days, express mail promises delivery within 24 or 48 h, and costs of about 20% more. For sellers and online stores, Ukrposhta also gave an opportunity to manage international shipments through their personal office, and for their customers—to inspect the physical goods in the department before paying for them. To streamline the company's navigation and improve customer communication, chat boots running on Facebook, Viber and Telegram platforms are launched [8].

Polish Post and Ukrposht do not perform those services that are offered by DHL in Poland, for example: comfort and flexibility of deliveries (last mile), market platform service (fulfillment) and supplies management. In addition, Ukrposht does not offer the following services: interactive messaging (predict), find a service point (pickup), personal collection at the facility (click and collect) and omnichannel. The service—common label SSCC (serial shipping container code) is offered by DHL in Germany and USA only.

7 Conclusions and Future Research

The chain of tomorrow's deliveries will be more compact, faster and, what is important, environment friendly and self-implementing. This unprecedented pace of changes will be driven by several state-of-the-art technologies, which will be cautiously accepted by the members of the industry in the coming a dozen or so years. The chain of deliveries currently constitutes a key competitive advantage for online trading companies. Such companies as Amazon on the European market and Zalando on the Polish market, which established new industry standards for innovative logistics in the year 2016, testify about the role of the delivery chain in the company's success. In the future, managers and financial teams will be putting more emphasis on the chain of deliveries and a reduction of costs. Approaches in the area of innovations are changing; micro-logistics is being introduced in the company with its impact on profits; macro-logistics is being introduced that has an impact on the growth of sale and customer loyalty. Those companies that sell on the Internet will have to evaluate their chains of deliveries and terminate obsolete and ineffective processes. Examples include everything: starting from the solutions of enterprise resource planning and electronic communication to forwarding agents and software for stock forecasting and management. Those IT companies that provide software focus on the transformation of efficiency, transparency and risk exposure to ensure advantage. Overall net savings can be measured by reduced data delays and faster response to changing trends. Retail chains will continue putting pressure on their suppliers to produce goods that are sold exclusively through their channels to be able to compete with and become different from such online sellers as Amazon. For those customers that show loyalty to these sellers, this will require greater specialization on the part of manufacturers without increasing shipping costs.

Automation and implementation of new technologies will help logistics enterprises ship goods to customers faster, at lower costs and with great precision; at the same time, however, the demand for warehouse workers will be diminishing. One of the best ways of achieving a significant reduction of shipping costs may be re-thinking of the design of the delivery chain as sustainable development. An expansion of the network of distribution and logistics centers makes it possible to position key resources closer to customers and shipping destinations. Placing the centres in adequate locations may reduce indirect shipping routes and zones, and this will finally contribute to reduced shipping costs and better environment protection. Owing to the delivery chain modelling, it is possible to find an adequate delivery chain design for the operator and postal services by experimenting with various scenarios and thereby reduce shipping costs. Changes in IoT will have a strong impact on the overall development of the network market (Networks), at the same time having an influence on solutions in the chain of deliveries. There will be an increased demand for comprehensive solutions in the chain of deliveries for mobile infrastructure networks including light assembly, local deliveries and spare parts management. At the same time, global level investments will differ as regards requirements related to the delivery chain, and emerging markets will require new network infrastructures and mature markets while focusing on an improvement of the network, thereby having an impact on their solutions as regards the delivery chain and sustainable development.

Polish Post and Ukrposht do not perform those services that are offered by DHL in Poland, for example: comfort and flexibility of deliveries (last mile), market platform service (fulfillment) and supplies management. In addition, Ukrposht does not offer the following services: interactive messaging (predict), find a service point (pickup), personal collection at the facility (click and collect) and omnichannel. The service—common label SSCC (serial shipping container code) is offered by DHL in Germany and USA only.

These days, logistics operators adapt their supply chains to suppliers and recipients to help them gain a competitive advantage. They form a SSN with them that is subject to constant changes. Smart supply chains play a key role in helping e-commerce companies to obtain the objectives of growth. On the present day markets, companies pose more and more frequently challenges to their supply chains on many levels, including flexibility, cost reduction, predictable risk management and customer service level, which is possible owing to the implementation of new technologies by logistics operators and also owing to a slow adaptation of national postal services.

The following research should be undertaken in the field of:

- whether and to what extent the new solutions affect customer service by the logistics operator?
- what pro-ecological elements influence the choice of a logistics operator in selected countries?
- whether the introduced elements of environmental policy influence the improvement of quality?

On the grounds of the analysis of data and the research in e-commerce market carried out, it can be stated that TSL companies demonstrate a sensitivity to activities connected with environmental protection, and they take responsibility for the effects of their negative impact. From the perspective of obtaining a competitive advantage, logistic operators are based on the creation of a positive image and an involvement not only as a company but also as its employees in social and eco-friendly activities. The introduction of ISO 14001 in logistics corporations resulted in procedural changes and having those subcontractors that realize transport within the framework of the logistic chain replace the means of transport with new and more eco-friendly means of transport.

References

1. Abukhader, S. M., & Jönson, G. (2004). Logistics and the environment: Is it an established subject? *International Journal of Logistics Research and Applications, 7*(2), 137–149.
2. Antonowicz, A. (2016). Handel internetowy—Implikacje dla logistyki (Internet commerce—Implications for logistics). *Handel Wewnętrzny, 2*(361), 8.
3. Bjartnes, R., Strandhagen, J. O., Dreyer, H., & Solem, K. (2008). Intelligent and demand driven manufacturing network control concepts. In *Proceedings of EurOMA/POMS Tokyo*, 5. October 8, 2008.
4. Chornopyska, N.V. (2012). Development of the express delivery market in Ukraine [Text]. In N. V. Chornopyska, N. R. Kubrak, & R. S. Chornopysky (Eds.), *Bulletin of the National University "Lviv Polytechnic". Logistics* (No. 749. pp. 310–315).
5. Defee, C. C., & Stank, T. P. (2005). Applying the strategy-structure-performance paradigm to the supply chain environment. *The International Journal of Logistics Management, 16*(1), 28–50.
6. Bereza, A. M., Kozak, I. A., Shevchenko, F. A., et al. (2002). *E-commerce: Tutorial manual* (p. 326). Kiev: KNEU.
7. Fichter, K. (2003). E-commerce. Sorting out of the environmental consequences. *Journal of Industrial Ecology, 6*(2), 29–30.
8. Grebenyk, K. (2017). The future of postal logistics in Ukraine and in the world: What market leaders are introducing today. Online https://mind.ua/publications/20177428-majbutne-poshtovoyi-logistiki-v-ukrayini-ta-sviti-shcho-lideri-rinku-vprovadzhuyut-uzhe-sogodni. Access: November 12, 2017.
9. Global B2C E-commerce Report. (2016). E-commerce foundation. *Amsterdam, 2017,* 11.
10. Heiskanen, E., Halme, M., Jalas, M., Karna, A., & Lovio, R. (2001). *Dematerilization: The potential of ICT and services.* The Finnish Environment 533. Helsinki: Finish Ministry of the Environment.
11. https://ecommerce.poczta-polska.pl/
12. http://www.dhl.com.pl/pl/logistics/
13. http://www.dhl.com.ua/uk/logistics.html
14. https://www.dhlparcel.pl/
15. https://www.envelo.pl/
16. https://www.minsait.com/en/what-we-do/empower/smart-supply-chain
17. http://www.poczta-polska.pl/
18. http://ukrposhta.ua/en/vidslidkuvati-forma-poshuku
19. Huang, Y. C., & Yang, M. L. (2014). Reverse logistics innovation, institutional pressures and performance. *Management Research Review, 37*(7), 615–641.
20. Kawa, A. (2014). Logistyka w obsłudze handlu elektronicznego (Logistics in the sernice of e-commerce). *Logistyka No, 4*(2014), 37.
21. Kim, W., & Mauborgne, R. (2015). *Blue ocean strategy: How to create uncontested market space and make the competition irrelevant* (Expanded ed.). Boston, MA: Harvard Business School.
22. Konopielka, Ł., Wołoszyn, M., & Wytrębowicz, J. (2016). *Handel elektroniczny. Ewolucja i perspektywy* (E-commerce. Evolution and perspectives) (p. 16). Oficyna Wydawnicza Uczelni Łazarskiego, Warszawa.
23. Kozerska, M. (2014). Obsługa logistyczna obszaru e-commerce (Logistic services for e-commerce area). *Zeszyty Naukowe Politechniki Śląskiej, seria Organizacja i Zarządzanie z., 68*(1905), 52.
24. Leszczyńska, A., & Maryniak, A. (2017). Sustainable supply chain—A review of research fields and a proposition of future exploration. *International Journal Sustainable Economy, 9* (2), 159–179.

25. Logistics 4.0 and smart supply chain management in Industry 4.0. https://www.i-scoop.eu/industry-4-0/supply-chain-management-scm-logistics/. Access January 31, 2018.
26. Macauley J., Buckalew, L., & Chung, G. (2015). *Internet Things in Logistics*. DHL Trend Research, Troisdorf, Germany, https://www.dpdhl.com/content/dam/dpdhl/presse/pdf/2015/DHLTrendReport_Internet_of_things.pdf. Access January 31, 2018.
27. Maevska, A. A. (2010). *Electronic commerce and law: Teaching methodical manual* (p. 256). Kharkiv.
28. Makarova, M. V. (2002). *E-commerce: A guide for students at Higher Educational Institutions* (p. 272). Kiev: Publishing Center "Academy".
29. Maryniak, A. (2017). Economic, environmental, marketing and logistic effects of intellectual capital resulting from the implementation of green supply chains, Вісник Національного університету "Львівська політехніка". *Логістика, 863,* 285–294.
30. Maryniak, A., & Strąk, Ł. (2017). Analyses of aspects of knowledge diffusion based on the example of the green supply chain. In N. T Nguyen, S. Tojo, L. M. Nguyen, & B. Trawiński (Eds.), *Intelligent Information and Database Systems, 9th Asian Conference, ACIIDS 2017, Proceedings, eBook, Kanazawa, Japan* (Part I, pp. 335–344), April 3–5, 2017.
31. McKinnon, A. (2010). Environmental sustainability: A new priority for logistics managers. In A. McKinnon, S. Cullinae, M. Browne, & A. Whiteing (Eds.), *Green logistics: Improving the environmental sustainability of logistics* (pp. 3–30). London: Kogan Page Limited.
32. Mgebrishvili Kh. A. (2016). Ecological aspects in logistics. In Kh. A. Mgebrishvili, N. V. Wutkhuzi, & H. A. Kvabelashvili (Eds.), *Modern technologies in engineering and transport* (Vol. 2, pp. 105–109).
33. Morgan, T. R., Richey, R. G., & Autry, C. W. (2016). Developing a reverse logistics competency the influence of collaboration and information technology. *International Journal of Physical Distribution & Logistics Management, 46*(3), 293–315.
34. Murphy, P. R., & Poist, R. F. (2000). Green logistics strategies: An analysis of usage patterns. *Transportation Journal, 40*(2), 5–16.
35. Osterwalder, A., Pigneur, Y., & Tucci, C. L. (2005). Clarifying business models: Origins, present and future of the concept. *Communications of the Association for Information Systems, 16*(1), 1–25.
36. Pluta-Zaremba, A. (2017). Rozwój usług logistycznych implikowany dynamicznym wzrostem rynku e-commerce (Development of logistic services implied by the dynamic growth of the e-commerce market). *Studia Ekonomiczne, Zeszyty Naukowe Uniwersytetu Ekonomicznego w Katowicach, No., 321,* 92.
37. Porter, M. E., & Millar, V. E. (1985). How information gives you competitive advantage. *Harvard Business Review, 63,* 149–160.
38. Report about e-commerce in Poland and East Europe. (2017). https://www.gemius.pl. Access December 27, 2017.
39. Roos, D. (2015). *The History of E-commerce*. HowStuffWorks.com. http://money.howstuffworks.com/history-e-commerce1.htm. Access April 26, 2015.
40. Sass-Staniszewska, P., & Binert, K. (2017). *E-commerce w Polsce 2016. Gemius dla e-commerce Polska* (E-commerce in Poland 2016. Gemius for Poland e-commerce). https://www.gemius.pl/wszystkie-artykuly-aktualnosci/nowy-raport-o-polskim-e-commerce-juz-dostepny.html. Access December 16, 2017.
41. Sokolenko, P. (2018). *Electronic commerce (e-commerce): trends and the forecast of development in Ukraine for 2017–2018.* https://www.web-mashina.com/web-blog/ecommerce-prognoz-elektronnoi-kommercii-ukrainy-2017-2018. Access January 10, 2018.
42. Stott, R. N., Stone, M., & Fae, J. (2016). Business models in the business-to-business and business-to-consumer worlds—What can each world learn from the other? *Journal of Business & Industrial Marketing, 31*(8), 949–950.
43. The Law of Ukraine "About e-commerce" on September 3, 2015, No. 675-VIII, as amended by the Law No. 1977-VIII of March 23, 2017, Data of Supreme Council of Ukraine, 2017, No. 20, p. 240.

44. Ułan, G. (2017). *Raport E-commerce w Polsce 2017—Prawie połowa polskich internautów nie kupuje online* (Ecommerce Report in Poland 2017— Almost half of Polish Internet users do not buy online). http://antyweb.pl/e-commerce-w-polsce-2017/. Access September 21, 2017.
45. Witthaut, M., Henning, D., Sprenger, P., Gadzhanov, & P., Dawid, M. (2017). Smart object and smart finance for supply chain management. *Logistics Journal.* https://www.logistics-journal.de/not-reviewed/2017/10/4610/witthaut_en_2017.pdf. Access January 31, 2018.
46. Wong, C. Y., Wong, C. W. Y., & Boon-itt, S. (2015). Integrating environmental management into supply chains. *International Journal of Physical Distribution & Logistics Management, 45*(1/2), 43–68.
47. Yuvaraj, S., & Sangeetha, M. (2016). *Smart supply chain management using internet if things (IoT) and low power wireless communication systems.* Wireless Communications Signal Processing and Networking (WiSPNET), International Conference, Chennai, Indie, March 23–25, 2016.
48. Zhu, Q., Sarkis, J., Lai, K. H., & Geng, Y. (2008). The role of organizational size in the adoption of green supply chain management practice in China. *Corporate Social Responsibility and Environmental Management, 15*(6), 322–337.
49. Zombirt, J. J., Bartoszewicz-Wnuk, A., Tęsiorowska, M., & Sosef, D. (2017). *Logistyka e-commerce w Polsce. Przetarte szlaki dla rozwoju sektora* (Logistics e-commerce in Poland. Wasted routes for the development of the sektor) (p. 11). http://www.jll.pl/poland/pl-pl/Research/Logistyka_e_commerce_w_Polsce_przetarte_szlaki_dla_rozwoju_sektora_raport.pdf. Access December 16, 2017.

Part II
Measurement and Improvement of Supply Networks

Smart Supply Network—Drivers, Opportunities, and Challenges

Meghdad Abbasian Fereidouni, Khatereh Azar Noor and Sara Ravan Ramzani

Abstract This paper details our effort in determining how technological advancement facilitates business relationship and ameliorates the competitive advantage of supply network. Using business network management literature, current knowledge employs Activity-Resource-Actor Model (ARAM) to recap the drivers, opportunities, and challenges of smart technologies in a supply network. The following discussions reveal that in presence of technological advancement, precise knowledge sharing, stronger social co-creation, smart management, and robust legal system, business relationship and supplier collaboration can benefit from a sustainable, modern, adaptive, robust, and technology-oriented activity links, resource ties, and actor bonds, per se, smart supply network. The paper confirms that digital-linked activities empower businesses and help them gain more from interdependencies benefits. Digital-linked resources enhance heterogeneity advantage, while digital-bonded actors are obtained from transcendence. The benefits derived from digitizing activity links, resource ties, and an actor bond is reliant upon a company's own activity structure, interdependencies, and connectedness. However, the challenges of direct/indirect relationship costs prevent companies benefiting from a smart supply network.

Keywords Business relationship · SMART supply network · Digital technology

M. A. Fereidouni (✉) · S. R. Ramzani
Center of Post Graduate Studies, Limkokwing University
of Creative Technology (LUCT), Cyberjaya, Malaysia
e-mail: mgabbasian@gmail.com

K. Azar Noor
Department of Economics, Multimedia University (MMU), Cyberjaya, Malaysia

© Springer International Publishing AG, part of Springer Nature 2019
A. Kawa and A. Maryniak (eds.), *SMART Supply Network*, EcoProduction,
https://doi.org/10.1007/978-3-319-91668-2_4

1 Introduction

Each part of today's supply chain has been digitized in order to facilitate interactions between stakeholders and supply networks. In the complex digital market no single actor could offer a service to customers on their own capability. Suppliers need a sustainable cooperation that creates a value network with the right partners. Partnership management competences must be a primary ability of new business actors. Digital technologies will also transform the actual structure of social relationships in both consumers and suppliers [32]. The services and products increasingly possess digital technologies (e.g. at the connected vehicle or smart house appliances), which is becoming harder to disentangle from the company processes using their primary IT infrastructures. In this context, Pagani and Pardo [24] point the requirement of designing a value chain while simultaneously engineering products/services and processes with the intention of creating value. Significant value could be created to assess the worthiness of appropriate knowledge dwelling at various points within the network, as well as arranging transfer to other points within the network.

Digital business methods are coordinating firms' product, procedure, and support domains, which creates complicated and powerful ecosystems [24]. The entire value network is actually supported by a particular worth, and its subsequent application leads to proper attitudes. Adopting the network perspective is more suitable for organizations, particularly those where its supply/demand chain is digitized. Some scholars [9, 18] specifically link a network's value to how is formed and based the idea of value network on "each product/service requires some value making activities carried out by numerous actors developing a value-creating system". Solberg et al. [26] define the value network as when a cluster associated with collaborating actors provide value towards the end customer and exactly where in the network each actor is responsible for. Based on this perspective, the value-creating system consists of different actors who interact to co-produce worth. Networks, as designs of actors, undertaking value activities, type the "environment" the actual firms tend to be embedded within. Understanding networks, their structures, procedures, and evolution is vital to its management. In same way, digitalization and advanced information and communication technology constantly reform the structures, procedures, and evaluation of network. Therefore, this research aims to find out how technology advancement and digitalization have been influencing business relationship and supply network.

2 Conceptual Backbone

Business network management research has been evolving since invention the Internet and development of advanced information and communication technology. As theoretical underpinning of business network management literature, a recent study from Möller and Halinen [19] summarizes the development of network management researches from 2000 to 2016 to six research streams and contributes General theory of networking management (NetFrame theory) as configuration of network management (influencing, orchestrating, and managing) in three layers of today network management (field layer, network layer, and actor layer). As special issues in Special Issue of Managing Business and Innovation Networks, the research figures out the strategic net perspective [20, 21] as enablers of network management that drawn on three types of strategic nets (current business nets, business renewal nets, and emerging business networks), management capabilities, and resource-based view and dynamic capability view. As second main research stream, strategizing in business network perspective [6, 13] concerns drawing limits to network management that dealing with IMP research versus strategy views, linking cognition and action, and limitation of managing in scope of action.

Möller and Halinen summarize cognitive perspective [10, 11] as third research stream in business network management that directing network management by working on network pictures and network theories, and sensemaking, visioning and construction. Knowledge perspective [3, 5] explains differences in network management by highlighting knowledge transfer, sharing and cocreation, different types of learning, and exploration versus exploitation. The fifth business network management research stream is called Institutional approach [22, 25] that extends and challenges network management by including institutional actors, highlighting collective action and changing institutions, and network influencing via stakeholder groups. As latest research stream, innovation networks [2, 7, 8] uncovers network orchestration by building on orchestration and actor types, radical versus incremental innovations, and innovation construction versus commercialization.

At this strand of research and based on upon Håkansson and Shenota [9], this study adapts the Activity Resource Actor Model (ARAM) to identify the effect of digitalization on the supply network. The ARAM not only uncovers activity-resource-actor relationships influenced by technology advancement, but it also provides the insight to single assessment of digitalization impact on activity, resource, and actor. The ARAM shows that a business exchange could be described in three "layers": Activity link, Resource ties, and Actor bonds. The model has the capacity to capture "the complex connections between activity coordination, resource combination, and actor link". ARAM views an activity like a "sequence associated with acts aimed towards the purpose". For example, developing the product, buying, selling, processing information is thought as activities. Activities could be raw provisions, physical amenities, components, products, in a nutshell, "various components, tangible or even intangible, materials or "symbolic", can be viewed as resources whenever use could be made associated with them". Within the same vein, actors connect others to mix resources and link activities. Actors within the ARAM model could be individuals or even organizations. The truth is that a company viewed as an Actor will be linked towards the idea that a company receives an identity with others (and not just because companies are thought of individuals, in a position to form intention, have reasons, or be a good agent).

In line with aforementioned "Activity", "Resource", "Actor", any B2B relationship can be evolved by new resources. Digital resources not only change the actor bonds, but it also causes new activity links. Thus, according to ARAM, digital technology brings digital-linked activity, digital-tied resources, and digital-bonded actors, and therefore, reforms the business relationship [24]. First, companies tend to be connected via Activity link, which issue technical, admin, commercial, and other Activities of the company that could be linked in multiple ways in order to those associated with another company. These types of connected Activities form a design. Companies will also be connected via Resource ties, which link various Resources. In a network concept, these linked Resources are a type of Resources pattern. Lastly businesses interconnect with other players via a web of actors in the network relationship.

These three associated layers are not independent, as there is relationship between them. However, the existence associated with bonds in between Actors is recognized as a prerequisite for the development associated with Activity link and Resource ties. The evolution of the business network could be described when it comes to changes affecting it if the pattern associated with activities and constellations associated with resources from the web are associated with actors [24, 29]. Activities could be changed via new changes and coordination. The Resource constellations could be modified whenever new mixture takes place, and the web associated with Actors is actually modified along with Actors altering their relationships. This paper focuses on how digital technology impacts activity links, resource ties, and actor bonds to highlight technology impact on the supply network and business relationships.

2.1   Digital-Linked Activity

Digital Resource can be used to enhance current activities by supporting a better (easiest, costless) coordination. This particular digitalization can be called the "Digital-linked Activity", since the primary impact from the digital Resource is about the links between Activities. Activities that are better coordinated due to digital technology could be "internal Activities", or even "external Activities" (Activities between two company Actors) [18, 19, 24]. An example from Pagani and Pardo [24] shows that a Digital Data Interchange (EDI) system does not fundamentally change the character of the actual activity between two Actors (the trade of information), it however allows it to work in far more efficiently. On the other hand, a Manufacturing Resource and Planning (MRP) system does not essentially change the operations of the company, however it allows a highly effective planning of necessary Resources (ibid). They also point how the term "does not really change essentially the activities" within the context where activities tend to be inevitably somewhat modified, in multiple ways, via digitalization. However, they cannot be looked at as "new Activities". Within the same research, it was mentioned that Biomérieux does not change the actual sale Activities, however, using electronic devices to accomplish the B2B dealings allows a far more efficient conversation. Also, the Coca-Cola Business, thanks to a digital CR platform, is able to continuously keep track of the customers and enhance its relationships (ibid).

2.2 Digital-Tied Resource

This kind of digitalization is principally characterized by a digital resource helping the development of brand new activities completed by existing actors. In this instance, it may be the combination from the digital Resources by an Actor using Resources associated with another that allows new Activities to look between them. This phenomenon results in the beginning of an digital ecosystems (different gamers collaborate to produce value). Connected objects can communicate towards the manufacturing company details about how they're used via customer businesses. On the foundation of these details, the provider can propose brand new services for the customers, such as the optimization of using products and instruction of providers. The framework reported by Pagani and Pardo [24] represents a brand-new Resource (provided by IBM and Dassault System) that allows the transformation of conventional business to produce new Activities. Dassault Systems works with different sectors to provide electronic solutions that integrate and alter current Activities (digitalized analysis tools, as well as augmented reality, which allows doctors to explore the heart, 3D publishing machines accustomed to improving efficiency in various businesses, and so on) (ibid).

2.3 Digital-Bonded Actor

In this type of transformation, the actual digital Resource facilitates new provision between Actors. This digitalization is called a "Digital-bonded Actor", since the primary effect of using electronic technology is the creation of new bonds between Actors via a new Actor going for a position within the network. If so, the electronic systems utilized by a new Actor permit connections between previously unlinked Actors or even modify them adequately to alter the character of actual bonding. Take for example a marketplace that uses actual digital Resource to permit buying/selling companies for satisfaction. This is often illustrated via a SpecialChem situation [24] that give the chemical substance company the chance to connect with similar businesses and enjoy knowledge sharing.

3 Drivers of Digital Supply Network

3.1 *Interdependency*

The various interdependencies tend to be intertwined between running a business activity and impacting business relationships. Technology driven modifications change the conditions and circumstances of interdependencies and consequently connectedness, which in turn, transform the value creation process and business relationships [9, 24]. Digital technology by supporting new activity links, making new resource ties, and empowering new actor bonds, facilitate supplier partnership. In this part, we assumed technology, knowledge, social relations, administrative routines and system, and legal ties as five drivers of Interdependency in supply network which along with connectedness are two main determinations of business relationship to examine digital technologies impact on new activities, resources, and actors in business relationship (ibid).

3.1.1 Technology

Companies within industrial marketplaces operate within a texture associated with available technologies. The specialized knowledge and technology being used are vital to business Activities [29]. The circulation of trade and relationships between two businesses reflects the actual technologies utilized by both. Connecting these systems is problematic, since Activities and adaptations are much more important relative to others. Similar to how a relationship evolves, possible specialized misfits need to be eschewed. Most of the adaptations produced in the businesses originated from the technical measurements of products and processes.

With an aggregated degree, technical interdependencies are seen as technological techniques, and in a few cases, known as a "digital platform" that offer a wide frame to business activity within industrial marketplaces [1]. These endeavors often embrace a number of stages associated with transformation and several sectors. The specialized connections shown in the associated platforms, as well as their development, are among major causes shaping the actual context of the company. The specialized connections help make relationships in a certain phase of transformation susceptible to, or the foundation of, changes within other occasionally rather distant regions of the technical system. Technical improvement within a company in its relationships would depend on the additional companies' systems; it is actually facilitated or even constrained not just by individuals with whom the organization maintains immediate relationships, but additionally by the actual technology associated with other third parties. Really, technical improvement often occurs within the actual frame associated with relationships to others. The specialized texture links different relationships to one another. Sometimes, we can easily see exactly how, for instance, the specialized development associated with equipment relates to the improvement of materials or the way the different items are used because components within the same finished product (system) should be related [4]. Moreover, situations pertaining to the actual role from the technical cable connections are much less evident, although it is of equivalent strength. The technology utilized by the parties to some business relationship has a tendency to influence not just the characteristics from the services and products exchanged, but the methods to do business, due to logistics, programs, planning, and management [30]. Business relationships are visible as hyperlinks that form and reflect current technology. The specialized connections between different relationships of the company, in many cases, are very powerful.

3.1.2 Knowledge

Every organization represents a mix of human and physical resources, which makes certain activities feasible. These tend to be tied to the activities of others. Beneath the actual activities of the industrial company, there's a pooling and combination from the knowledge, as well as skills from individuals [18]. The actual tacit understanding, which is essentially the combined understanding of those getting involved in a company, is typically regarded as its main property. When various company activities are completed and resources are utilized, some type of combination knowledge is required. This understanding of resource use is just partly specific, which implies that it may be articulated, codified as documents or even books, and is therefore relatively simple to move. Perhaps the primary main knowledge necessary to use resources and undertake activities is harder to outline. It is actually 'tacit' within nature, meaning that it is harder to move, as it is unique to individuals and is dependent on and developed using their respective past encounters [27].

The understanding of the organization reflects not just the information of its personnel, but that from other businesses and those it is connected to via business relationships [28]. The height of knowledge in a company is made available via its relationships with others. The Activities of the company are therefore made feasible by the knowledge of a few people. It becomes obtainable via relationships with customers and suppliers with others. It could be activated and operated when necessary. As the actual relevant understanding is spread among various actors (other companies) within the company, the use of some with no counterpart implies that only specific knowledge could be used (ibid). The understanding of a company and its competence would depend on its relationships, which are essential tools that connect the data of numerous different actors.

Since the competence of the company would be to a large extent reliant on its relationships, the improvement of understanding (the improvement of competence) would be, to a large degree, occurring in individual relationships [15]. The relationships are how we understand other parties and forms of knowledge. As this new understanding is generally linked to both sides, it means that the data of both parties are linked. This method of understanding development continues with a number of parties the organization has for connecting and incorporating pieces. From an understanding perspective, the company could be perceived as a unit that includes different bits of knowledge. The effect of understanding connections about the competence, and therefore overall performance of businesses is powerful, as demonstrated by, for instance, the need for different 'industrial districts' or even local networks.

3.1.3 Social Relations

Business relationships tend to be handled by individuals with different interpersonal roles. Social provisions that occur among individuals within two companies are essential for shared trust as well as confidence [17]. People interacting and representing their respective organizations within a business relationship undertake other functions in additional contexts. They are a part of other relationships; fit in with professional organizations, are family members, neighbors, or even schoolmates, have possibly developed other forms of individual relationships within other circles, created numerous social provides in operating places, interpersonal and wearing clubs, religious organizations, and others. Every person's social network is made up of personal relationships. It may be used in many ways to enhance or develop company relationships. These types of personal networks may, within well-established commercial networks, be of the 'clan' structure [9]. The expert networks within a certain industry are definitely an example. They make it difficult for any person missing the 'right' history and connections from being accepted and perform successfully.

3.1.4 Administrative Routines and Systems

A large amount of what's going on in a relationship is actually administrative by nature [9, 14]. It will find rules as well as norms within a context of the business that enforce activities; meetings tend to be held and papers and documents 'processed' to adhere to business exercise. There tend to be other responsibilities imposed upon companies via legislation. The majority of the administrative activity is some type of information digestion/control that is necessary to facilitate actual coordination associated with behavior amongst different events. Information digestion and trade communication running a business is extensive and costly. Within buyer-seller relationships, various attempts tend to be designed to improve the actual efficiency from the information digestion by creating rules and administrative programs. Some businesses develop information systems, often typical to numerous companies, to handle expenses and problems associated with information [24]. There has been an attempt to create general types of communication techniques in multiple categories of companies and industries. Development associated with industry requirements and norms is actually another substantial factor in this context.

The options adopted in a single (or several) relationship(s) could affect what's possible or essential in other relationships [18]. In the event that, for instance, a supplier really wants to sell to some large vehicle manufacturer, it would probably need to join the supplier's info system. This can, in effect, change what it could do for other clients. It is going to be easier for that supplier to serve other clients utilizing the exact same system compared to customers utilizing another program. The also applies to industry requirements and norms. This is how the admin systems create relationships. Entering the booking system of an airline company connects the tour owner with air-carriers. Selling to a nuclear energy equipment producer requires conformity with numerous quality guarantee routines and rituals. Connections, along with important outcomes, could therefore exist between different facets of the company due to the administrative programs and techniques.

3.1.5 Legal Ties

Besides the greater general program of guidelines and norms, the texture associated with control (influence) could be labelled due to legal consistency often being contained in the framework of company organizations [18, 24]. The legal consistency is associated with interest, as it can connect different sections that are similar. This is applicable, especially to different types of ownership management/other styles of contracts. There might be ownership ties that take multiple forms. Some international companies tend to be organized in many impartial companies that have developed internally by creating new models in international countries/ technologies. Other companies fit in with pretty much extensive conglomerates where the mutual trade relationships tend to be weaker, but seldom minor. Priorities may be given, officially/informally, to purchase from and sell to companies linked to them. Other kinds of legal interdependencies would be the different official cooperation agreements of numerous types via joint endeavors, to licensing contracts. An additional example is the procurement guidelines, common in several fields associated with international companies, that enforce some extent of 'local content' within supplies and other similar legal needs [23]. In a few industries, legal ties in certain form are common, as they ensure relationships between suppliers, customers, and third parties within a company remain interdependent. The numerous interdependencies could be used to expand or curtail the company connections. They are also applicable in a permanent, short term, or long-term objectives by the people, companies, or even departments. They can be consciously exploited by companies' due to its inherent advantages vis-à-vis relationships.

3.2 Connectedness

In the context of different interdependencies associated with business relationships as well as their impact on a relationship, the problem of connectedness is being in contact. The idea of interdependence associated with business relationships is generically applicable; things happening inside a relationship have bearing on other relationships [18, 24]. The universal interdependence associated with business relationships is rather obvious. What a business can offer within a relationship for its customers depends upon its relationships with suppliers. But there is more to this relationship between corporations. 'Connectedness' is all about these connections; relationships tend to be connected whenever a given relationship impacts or suffers from what's going on in other relationships. Not every relationship is connected. What goes on in the relationship to some customer may, for instance, affect what's happening with the relationship with another client. A change inside a relationship between a company and a supplier could positively/negatively affect customer relationship.

Connectedness associated with specific relationships of a company is usually recognized as well as held within business relationships [18]. How handled connections tend to matter significantly for relationship development and the actual performance of the company. The type and quantity of Resources a business can access affect its capacity to confirm a relationship. For instance, a corporation's relationship with its main customer, connections, other clients, and suppliers in relation to consultants, banking institutions, or investigation institutions are available. It is also obvious how these relationships can be cultivated for other relationships, but its effects could also be negative. An item developed with the customer could be of benefit to some other clients with similar needs, but not others. However, it still requires essential development resources. Similarly, the development of a certain administration routine within the relation using the customer could be positive towards additional customer relationship, but it makes it difficult to respond successfully to customers with different requirements (ibid). The relationship with the customer is actually handled by people. Social provisions are developed and essential, and applicable between relationships. The most common method to determine the suitability of a new partner is via referrals, i.e. looking into how it handles previous relationships [4]. The best references refer to an individual already recognized as an equal. Personal connections, in many cases, are a tool in trust-building. A legal contract or possession link is visible as both a benefit and disadvantage. It may also be regarded as a threat when a customer is really a competitor, but it could also be viewed as strong if it is a contrasting producer.

Connections between one customer relationships to another are relatively simple to determine, although it is much less obvious. However, it does not mean that it is less important. The need to connect to a client is evident when a company buys, as inputs, more than 50% of its turnover. To be able to succeed in a client relationship, support by suppliers is required. Technical cooperation needing a supplier is vital towards customer relationship. The chance offering Just-in- time (JIT) deliveries towards a customer could rely on a particular supplier's capability to deliver on time [12]. A supplier's knowledge may well be advanced to adapt to products for a customer. Quality assurance by a provider could mean more business. Similarly, having a relationship with some horizontal unit could be important towards customer relationship. The horizontal units that could affect customer relationship tend to be numerous: banking institutions, owners, and attorneys, international committees within standardization or even trade, and so forth. A company is an entity that is able to create capabilities and strength required to perform within a relationship and form new ones. This could result in its improvement.

In summarizing interdependencies and connectedness of actor relationship, Fig. 1 demonstrates digital-enabled supply network ecosystem comprises of three layers of digital-tied resources, digital-linked activity, and digital-bonded actors. This framework recaps the above-mentioned drivers of smart technology in supply network according to social relations, knowledge, technology, legal ties, and administrative routines and system. Unlike simple reciprocal relationship drawn in contribution of other scholars [18, 24], the digital-enabled supply network ecosystem shows complex, novel, embedded, and dynamic network environment [19] between actors and their activities and resources. However, framing from ARAM concept, digital-enabled supply network ecosystem illustrates the actor relationship with other actors via digital-linked activities (e.g. social campaigning, information sharing, electronic commerce, and electronic governance) while sharing digital-tied resources (big data, Internet of Things (IoTs), and digital platform).

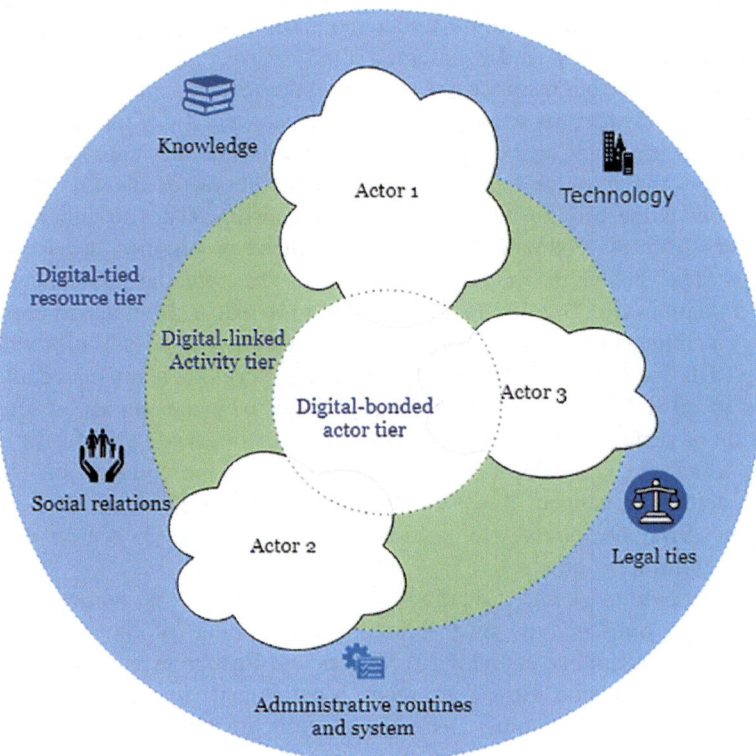

Fig. 1 Digital-enabled supply network ecosystem (authors' proposed framework)

4 Challenges and Opportunities

4.1 Relationship Opportunities

It was pointed out earlier that benefit originate from taking advantage of interdependencies within activity design, the heterogeneity within resource constellation, and transcendence within networks associated with actors. Activity links take advantage of the interdependencies, Resource ties make the most of heterogeneity, and Actor provision take advantage of transcendence [18, 19, 24]. Nevertheless, the advantages of activity links are determined by how extensive a particular activity is with regards to the company's personal activity framework. The bigger the reveal is from the company, the bigger the possible economic advantages of exploiting the present interdependencies. The possibility of cost savings via a better linking ought to be pretty much directly proportional towards the 'size' from the activity (ibid). Furthermore, activities that are interdependent in a number of dimensions may have bigger potential advantages than activities along with fewer or even more marginal interdependencies. Moreover, if the actual Activity framework is so that the interdependencies can certainly be recognized and dealt with, the advantages of linking increases. Thus, the interdependencies between Activities could be exploited, depending on a family member and absolute amount of the Activities, the level of interdependency, and its structure.

The advantages from resource ties stem from the way the relationship is retained. They're used in order to exploit the actual heterogeneity within the resources. Resource ties are going to be especially important once the heterogeneity within resources assortment of the organization is higher. One factor affecting the value of resource ties is the way an item within the relationship relates to either the actual capabilities from the company to Resources utilized or supplied in its other relationships. The greater the interfaces, the bigger the potential advantages of ties within a relationship. The 2nd factor is linked to its equal. If an actual counterpart employs a composite group of Resources again, then larger possibilities will create new ties. The 3rd factor is actually how a Resource selection is organized affects the options to advantage via this type of ties. To convey the impact of situational factors on advantages of Resource ties the worth of ties having a specific equal is biggest when: (1) the group of Resources utilized by the equal is amalgamated; (2) the actual exchanged Resource component is closely associated with the capabilities from the company to its additional relationships; (3) you can easily link actual relationship to others.

In terms of Actor sizing, a relationship produces bonds that works as a type of mechanism to transcend its own limitations. The value of those bonds would depend on the possibility of linking Activities and Resources, and the alternatives perceived for every side. Bonds are really a prerequisite in terms of creating and exploiting actual opportunities associated with linking. They'll consequently become more valuable once its potential is large enough. When the counterpart offers some options, providers become much less valuable. However, when an organization has a few alternatives, actual bonds tend to be more valuable. To convey the hypothesized relationships; bonds having a specific counterpart tend to be more valuable whenever: (1) the actual potential results of hyperlinks and scarves are considerable; (2) the actual counterpart offers many options; (3) the organization has a few alternatives.

Analyzing potential effects from the three types of substance within the relationships of participating businesses resulted in impacting factors. Advantages of a relationship tend to be situation-specific. The discussion [31] suggests that they will be dependent upon three kinds of factors: the trade content from the relationship, the options that come with the counterpart and people of the organization itself. The potential advantages of a relationship for any company tend to suffer from two factors associated with the trade content from the relationship [18]. One is the amount, while the other is the dependency between our item and activities, resources, or relationships from the company. Two factors stem from the characteristics counterpart. One factor has to do with how complicated the group of Resources utilized by the equal is, while the other is the number of alternatives the actual counterpart offers. Finally, two factors have been identified within an organization: the number of alternative counterparts the organizations hand how it organizes activity framework and resource selection, i.e. if we can easily connect links and ties within a relationship to others.

4.2   Relationship Challenges

Various difficulties arise when examining the effect of price on relationship. Some include traditional accounting difficulties of attribution and allocation associated with expenses, while others are due to the relationship Connectedness and the real cost results from different times. In classifying relationship based on costs, the conventional distinction between direct/indirect expenses can be used [9]. The direct costs include all costs that may be directly traced to a single relationship. Examples include the time invested in product sales or support contacts, transport, and customer-specific modifications. Most costs involve monitoring an actual relationship and building and maintaining it. There tend to be, however, additional expenses, which are collectively called the indirect costs. These could be bottom costs and can include expenses covering inner activities, which are a required condition to maintain a particular relationship heading (for instance, production, storing, and development from products exchanged within the relationship), but which are also used by other relationships [1]. Therefore, a distinction could be made between two types of internal Activities; those that directly could be traced to some certain relationship, and people which cannot be traced. The relationship expenses are closely associated. Increased dealing with costs may decrease bottom costs (adaptations resulting in increased linking), and increased bottom costs (investments within flexible machines) could also reduce the actual handling costs in a single particular relationship [4].

An additional taxonomy foundation that seems appropriate is the actual distinction between costs associated with day-to-day activities and investment expenses [31]. Every relationship leads to day-to-day expenses, but it's also possible for every party in a relationship to purchase technical, as well as organizational facilities to be able to decrease actual day-to-day expenses. Combining both dimensions provides a matrix, along with four types of costs incurred due to a relationship for a company. Each category offers some peculiarities, but there're also important trade-offs. This will connect more individuals and result in increased adaptations. Companies will also become much more involved, which could in theory result in decreased bottom costs. This relationship can be improved and Resources could be easily tied.

In terms of the total expense of relationship opportunities and day-to-day costs, the results vary. The 'larger' the actual relationship is when it comes to exchange, the greater the relationship investments may become [16]. System activities may take over random activities, which will generally decrease dealing with cost for each unit as long as it proceeds. The costs of the relationship increased since it came into play. Relationships offering exactly the same benefits that are developed towards the same degree are still prone to various costs. One counterpart could be less suitable, and therefore more expensive.

5   Conclusion and Theoretical Implications

This study rationally summarized the drivers, challenges, opportunities of technology in boosting business relationship, and ameliorating supply networks in a B2B organization. The research initially aimed to investigate how technology impact business relationship and suppliers network by employing an in-depth review of business relationship literature in service and manufacturing research and utilizing theoretical underpinning of supply network studies. The ARAM (Activity-Resource-Actor Model) reveals that the interdependencies and connectedness in business are reinforced by digitalization impacts via technology, knowledge, social relations, administrative routines and system, and legal ties. In essence, precise knowledge sharing, stronger social co-creation, smart management, and robust legal system, business relationship and supplier collaboration benefit from sustainable, modern, adaptive, robust, and technology oriented activity links, resource ties, and actor bonds, per se, and SMART supply network.

It has been ascertained that activity links empower business to gain the interdependencies benefits, resource ties enhance heterogeneity advantage and actor bonds accomplished from transcendence. The benefits from digitization of activity links, resource ties, and actor bonds depended on the companies' own activity structure, interdependencies, and connectedness. However, the challenges of direct/indirect relationship costs prevent the companies benefiting from the smart supply network. Digital technology change business environment, network component, actors by transforming resource integration, activity innovation and actor's role in business relationship and supply network. Value comes forward via a process associated with activities justification or innovation, depending on new activities emerging due to new digital-linked resources or innovation because new actors conduct activities.

Notwithstanding natural restrictions of conceptual researches, this study has some limitations. Due to broad implication of ARAM in assessing digitalization effect on business networking, the review-based qualitative methodology employed in this paper can be broadly challenged in regard to its simplicity and opacity. The further conceptualization can be more rigid and thus focused on other networking theories that bring specific results. Moreover, lack of empirical research limits the results associated with this conceptual paper. Therefore, future research can tests the proposed framework with real case study and evaluate the results respectively.

References

1. Aarikka-Stenroos, L., & Ritala, P. (2017). Network management in the era of ecosystems: Systematic review and management framework. *Industrial Marketing Management, 67*(April 2016), 23–36. https://doi.org/10.1016/j.indmarman.2017.08.010.
2. Adner, R., & Kapoor, R. (2010). Value creation in innovation ecosystems: How the structure of technological interdependence affects firm performance in new technology generations. *Strategic Management Journal, 31*(3), 306–333.
3. Araujo, L. (1998). Knowing and learning as networking. *Management Learning, 29*(3), 317–336.
4. Chen, L., Zhao, X., Tang, O., Price, L., Zhang, S., & Zhu, W. (2017). Supply chain collaboration for sustainability: A literature review and future research agenda. *International Journal of Production Economics, 194*(April 2016), 73–87. https://doi.org/10.1016/j.ijpe.2017.04.005.
5. Dyer, J. H., & Nobeoka, K. (2000). Creating and managing a high-performance knowledge-sharing network: The Toyota case. *Strategic Management Journal, 21*(3), 345–367.
6. Gadde, L.-E., Huemer, L., & Håkansson, H. (2003). Strategizing in industrial networks. *Industrial Marketing Management, 32*(5), 357–364.
7. Håkansson, H. (1987). *Industrial technological development: An interaction approach.* London, United Kingdom: Croom Helm.
8. Håkansson, H. (1990). Technological collaboration in industrial networks. *European Management Journal, 8*(3), 371–379.
9. Håkansson, H., & Shenota, I. (1995). *Developing relationships in business networks.* London: Routledge.
10. Henneberg, S. C., Mouzas, S., & Naudé, P. (2006). Network pictures: Concepts and representations. *European Journal of Marketing, 40*(3/4), 408–429.
11. Henneberg, S. C., Naudé, P., & Mouzas, S. (2010). Sense-making and management in business networks—some observations, Considerations, and a research agenda. *Industrial Marketing Management, 39*(3), 355–360.
12. Ho, H. D., & Lu, R. (2015). Performance implications of marketing exploitation and exploration: Moderating role of supplier collaboration. *Journal of Business Research, 68*(5), 1026–1034. https://doi.org/10.1016/j.jbusres.2014.10.004.
13. Holmen, E., & Pedersen, A. (2003). Strategising through analysing and influencing the network horizon. *Industrial Marketing Management, 32*(5), 409–418.
14. Jain, V., & Benyoucef, L. (2008). Managing long supply chain networks: Some emerging issues and challenges. *Journal of Manufacturing Technology Management, 19*(4), 469–496. http://www.emeraldinsight.com/doi/10.1108/17410380810869923.

15. La Rocca, A., Perna, A. Snehota, I., & Ciabuschi, F. (2017). The role of supplier relationships in the development of new business ventures. *Industrial Marketing Management* (December 2015), 1–11. http://dx.doi.org/10.1016/j.indmarman.2017.12.008.
16. Makkonen, H., Vuori, M., & Puranen, M. (2016). Buyer attractiveness as a catalyst for buyer-supplier relationshipdevelopment. *Industrial Marketing Management, 55,* 156–168. https://doi.org/10.1016/j.indmarman.2015.09.004.
17. Mani, V., Gunasekaran, A., & Delgado, C. (2018). Enhancing supply chain performance through supplier social sustainability: An emerging economy perspective. *International Journal of Production Economics, 195*(October 2017), 259–272. https://doi.org/10.1016/j.ijpe.2017.10.025.
18. Möller, K. K., & Halinen, A. (1999). Business relationships and networks. *Industrial Marketing Management, 28*(5), 413–427. http://linkinghub.elsevier.com/retrieve/pii/S0019850199000863.
19. Möller, K., & Halinen, A. (2017). Managing business and innovation networks—from strategic nets to business fields and ecosystems. *Industrial Marketing Management, 67* (November), 5–22. https://doi.org/10.1016/j.indmarman.2017.09.018.
20. Möller, K., & Rajala, A. (2007). Rise of strategic nets—new modes of value creation. *Industrial Marketing Management, 36*(7), 895–908.
21. Möller, K., & Svahn, S. (2003). Managing strategic nets: A capability perspective. *Marketing Theory, 3*(2), 201–226.
22. North, D. C. (1991). *Institutions, institutional change and economic performance.* Cambridge: Cambridge University Press.
23. Nuroğlu, H. H. (2016). Business network governance structure and IT capabilities. *Procedia—Social and Behavioral Sciences, 229,* 50–59. http://linkinghub.elsevier.com/retrieve/pii/S1877042816310485.
24. Pagani, M., & Pardo, C. (2017). The impact of digital technology on relationships in a business network. *Industrial Marketing Management, 67*(August), 185–192. https://doi.org/10.1016/j.indmarman.2017.08.009.
25. Scott, W. R. (2001). *Institutions and organizations.* Thousand Oaks, CA: Sage.
26. Solberg, K., Aracena, M., and Jallouli, R. (2012). Key success factors for ericsson mobile platforms using the value grid model ✩. *Journal of Business Research, 65*(9):1335–45. https://doi.org/10.1016/j.jbusres.2011.10.002.
27. Syntetos, A. A., Babai, Z., Boylan, J. E., Kolassa, S., & Nikolopoulos, K. (2016). Supply chain forecasting: Theory, practice, their gap and the future. *European Journal of Operational Research, 252*(1), 1–26.
28. Wang, Y., Wallace, S. W., Shen, B., & Choi, T. M. (2015). Service supply chain management: A review of operational models. *European Journal of Operational Research, 247*(3), 685–698. https://doi.org/10.1016/j.ejor.2015.05.053.
29. Wycisk, C., McKelvey, B., & Hülsmann, M. (2008). 'Smart parts' supply networks as complex adaptive systems: Analysis and implications. *International Journal of Physical Distribution & Logistics Management, 38*(2), 108–125. http://www.emeraldinsight.com/doi/10.1108/09600030810861198.
30. Yan, T., Yang, S., & Dooley, K. (2017). A theory of supplier network-based innovation value. *Journal of Purchasing and Supply Management, 23*(3), 153–162. https://doi.org/10.1016/j.pursup.2017.02.002.
31. Yang, Z., Zhang, H., & Xie, E. (2017). Relative buyer-supplier relational strength and supplier's information sharing with the buyer. *Journal of Business Research, 78,* 303–313. https://doi.org/10.1016/j.jbusres.2016.12.026.
32. Zhong, Ray Y., Newman, S. T., Huang, G. Q., & Lan, S. (2016). Big data for supply chain management in the service and manufacturing sectors: Challenges, opportunities, and future perspectives. *Computers & Industrial Engineering, 101,* 572–591. https://doi.org/10.1016/j.cie.2016.07.013.

Measuring Performance of Adaptive Supply Chains

Dorota Leończuk, Urszula Ryciuk, Maciej Szymczak
and Joanicjusz Nazarko

Abstract In the wake of the intensification of competitive struggle that we can call hyper-competition and in the face of temporary, transient and often unsustainable competitive advantage supply chains have to master their processes in many dimensions at the same time. Excellence is achieved through a shared vision of development and cooperation with up and down-tier supply chain members especially by continuous assessment and improvement the effectiveness and efficiency of the supply chain processes. Typical determinants of the supply chain performance is the triad: level of customer service—time—costs. However, intensive changes taking place in the supply chains surroundings enforce the inclusion of new criteria in supply chain performance measurement. In the chapter the problem of supply chain performance measurement with reference to the concept of adaptive supply chains was considered. The study was based on quantitative research conducted among Polish companies employing 50 or more employees from four sectors of economy: automotive, food, furniture as well as consumer electronics and household appliances. 200 computer assisted telephone interviews (CATI) were held. According to the conducted research the scale for measuring the supply chain performance should take into account four factors, namely responsiveness, versatility, velocity, and visibility (3V + R formula).

D. Leończuk · U. Ryciuk · J. Nazarko
Faculty of Engineering Management, International Chinese
and Central-Eastern European Institute of Logistics and Service Science,
Bialystok University of Technology, Bialystok, Poland
e-mail: d.leonczuk@pb.edu.pl

U. Ryciuk
e-mail: u.ryciuk@pb.edu.pl

J. Nazarko
e-mail: j.nazarko@pb.edu.pl

M. Szymczak (✉)
Faculty of International Business and Economics, Department of International Logistics,
Poznań University of Economics and Business, Poznań, Poland
e-mail: maciej.szymczak@ue.poznan.pl

© Springer International Publishing AG, part of Springer Nature 2019   89
A. Kawa and A. Maryniak (eds.), *SMART Supply Network*, EcoProduction,
https://doi.org/10.1007/978-3-319-91668-2_5

Keywords Supply chain performance measurement · Adaptive supply chain 3V formula · Smart supply chain · Exploratory factor analysis (EFA) Confirmatory factor analysis (CFA)

1 Introduction

Nowadays supply chains need to keep pace with competition to more extent than ever before. In the wake of the intensification of competitive struggle that we can call hyper-competition [15] and in the face of temporary, transient [45] and often unsustainable competitive advantage [44] supply chains have to master their processes in many dimensions at the same time. It is said that they have to raise their level of excellence, they have to be managed towards maturity. Maturity itself is the term that has already been well received in the context of increasing supply chain performance. Achieving supply chain maturity is a strategic task that requires co-operation of all entities involved. Only then it is possible to increase process capability, effectiveness and productivity. Supply chain maturity models assume the existence of different levels on which supply chains can be placed. Supply chain maturity can be determined by control (the difference between targets and actual results), predictability (the variability in achieving objectives) and effectiveness (the achievement of targeted results and the ability to raise targets) [43]. The most widespread supply chain maturity models are:

- the "compass" model [60],
- PRTM/PMG model [3],
- and Poirier's model [53].

The order in which the three models are mentioned here corresponds to the chronology of their development. All of them were published at the beginning of the 21st century not counting the very first preliminary version of the Poirier's model. Each of them shows the achievement of supply chain maturity in a very similar way. They exhibit only a certain shift of focus. Maturity (excellence) is achieved through a shared vision of development and cooperation with up and down-tier supply chain members especially by continuous assessment and improvement the effectiveness and efficiency of the supply chain processes.

Typical determinants of the supply chain performance is the triad: level of customer service-time-costs. However, intensive changes taking place in the supply chains surroundings enforce the inclusion of new criteria in performance measurement and improvement, for example the ability to react quickly or to operate flexibly. Main factors influencing supply chains evolution are related to globalisation, increased customer expectations as well as technologisation and technicisation of life [62].

Today's supply chains should be characterised mainly by visibility, velocity and versatility. It is so-called "3V rule" [56]. Supply chains outstanding in the characteristics that could be described by the 3V formula parameters are adaptive supply chains (having the features of flexible, responsive and resilient). The new concept are also smart supply chains. Smart "is the coming together of software, hardware, cloud and sensing technologies so as to be able to capture and communicate real time sensor data of the physical world, which can be used for advanced analytics and intelligent decision making" [47]. However, smart supply chains are also integrated, intelligent and innovative [74]. Smart supply chains with main distinction of adapting to technological changes and its consequences could be also called adaptive.

Managing an adaptive supply chain, maintaining and developing all its capabilities requires a system that will show which way to follow. Performance measurement systems fulfil this role. Performance measurement includes a usage of set of diagnostic tools used to measure, monitor and assess the processes in supply chain. That's why they are currently so popular and why hereby, in the chapter the problem of supply chain performance measurement (SCPM) with reference to the concept of adaptive supply chains was considered. The purpose of the study was to indicate the factors that should be taken into account when measuring supply chain performance in order for the supply chain to gain the feature of adaptability and show its excellence in this regard as it develops in time. The rationale for doing such research was to verify and confront with the approach referred to as the 3V formula.

The study was based on quantitative research conducted among Polish companies employing 50 or more employees from four sectors of economy: automotive, food, furniture as well as consumer electronics and household appliances—sectors that are indisputable leaders of Polish export. 200 computer assisted telephone interviews (CATI) were held. Interviews were conducted with the management staff on the basis of a structured questionnaire. The research was preceded by a comprehensive literature review that allowed for identification and selection of indicators of supply chain performance.

2 Smart and Adaptive Supply Chains

Adaptability is one of the most significant features of a supply chain that has an impact on the results of its functioning. The aptitude of a supply chain in terms of adjusting to all the more challenging operational conditions is one of the paramount factors that guarantee long-term competitiveness and success. Ivanov et al. [30] claim that a supply chain can be called adaptive if it is capable of adapting to:

- changes in the market environment and the functioning in conditions of uncertainty,
- changes in the executive environment of specific measures,
- internal changes in the supply chain itself

by means of using structural and functional reserves as well as better coordination that results from the application of information and computer technologies, in particular the Internet. Under the influence of long-term and strong changes in the environment, this type of supply chain is able to reduce, suppress or eliminate disruptions and maintain, or even improve the operational efficiency through reconfiguring its elements (transition to a new state).

The adaptive capacity is an effect of developing a certain set of features in the supply chain. Among the major features in the supply chain many authors indicate visibility, velocity and variability, known as the 3V's [31, 56]. Kalakota et al. [37] consider inventory visibility, fulfillment velocity and coordination versatility the three fundamental pillars of adaptive supply chains.

Mastering many supply chain processes at the same time means that the supply chain can stand out in a variety of areas simultaneously, can exhibit many different characteristics, can have many abilities and maintain a competitive edge in many different areas. One may call such a supply chain a smart one. Butner [11] states that supply chains must become a lot smarter to deal effectively with risk and meet business objectives. She argues that we can expect a different kind of supply chain to emerge that is instrumented ("full of sensors, RFID tags, meters, GPSs, and other devices and systems"), interconnected (characterised by "unprecedented levels of interaction with customers, suppliers, and IT systems in general, but also among objects that are monitoring or even flowing through the supply chain") and intelligent (capable of learning and making some decisions by itself, without human involvement as well as to predict future scenarios) [11]. Other researchers go even further. For example, according to Wu et al. [74] a smart supply chain possesses as many as six distinctive characteristics—it is instrumented, interconnected, intelligent, automated, integrated, and innovative.

The concept of smart supply chain derives from customers' expectations of high quality service adjusted to fast-changing needs. Now it is oriented to adding value to customer. To imagine this phenomenon a term 'smart convenience" has been coined. How to be smart convenient in supplying goods is particularly visible when selling foodstuffs. Convenient food shopping has progressed considerably from convenience stores through strip malls, leisure-oriented and personalised supermarket towards combining traditional shopping done in a traditional store with on-line shopping [63]. This concept is now the mainstream of smart convenience.

With the development of online sales, mixed shopping experience offering ordering online and the convenience of being able to pick up your order while you're out and about, and omnichannel sales, distribution channels will have to evolve as well as supply chains will have to operate under new schemes and business models [24]. Yet another incentive to develop smart supply chains is factory-on-demand model within dispersed manufacturing networks [49] which is strictly connected to the above mentioned customers' expectations but considered from the other side of distribution channels.

Smart supply chain is an integrated supply chain as the integration is the only solution providing high levels of flexibility, what enables the supply chain to respond immediately to demand signals with the ability to communicate and collaborate with each other and trading partners. Smart supply chains should be also featured by high level of trust between supply chain members as well as trust between information systems and information technologies. The most mature supply chains ensure real-time information and real collaboration between partners. Smart technologies introduced into supply chains could lead to real time transparency and higher levels of visibility. "The transparency within the value chain allows the manufacturer to identify changes in customer requirements and to reflect them in all of the production steps, from development to distribution" [66]. The visibility especially relates to inventories and possibility to monitor its levels anytime and from anywhere in the chain.

Smart supply chain is also resilient supply chain adapting strategies and operations to changes in the environment to reduce the risk of activity. It means actions are aimed at avoiding disturbances or reducing severity if they occur [8]. In the future, smart supply chains will evolve into self-adaptive supply chains where all things will be interconnected, exchanging information, recognizing and assessing situations. To collect/store/analyse the information in real time, IoT technologies, Radio Frequency Identification, sensors, GPS and BigData (including distributed databases and distributed parallel processing frameworks) are essential [38]. The supply chain of the future will fully rely on digital solutions and will be self-driving powered by artificial intelligence. An intelligent supply chain will automatically balance "supply with demand, with demand forecasting systems based on historical forecasts and predictive algorithms" [68].

3   Supply Chain Performance Measurement

Performance measurement includes a set of diagnostic tools used to measure, monitor and assess the effects of an organisation, which is an essential element of internal management control systems. It is a closed loop system that closely monitors the effects at the operational, tactical and strategic level of management, and thus operate in an efficient and effective manner [7]. Keeping up with the needs of consumer, more and more frequent changes in product range with the ever-increasing volumes of data, make changes in the field of business analytics. Analytical procedures need to be implemented at every stage of the supply chain and are focused on predicting future events so that the supply chain can be managed proactively. Today we've got a variety of data sources that have emerged in recent years. There should be mentioned various types of social networks operating in real time, which are not only a source of data about customers, but also serve as an internal organisational communication platform (e.g. Yammer, Chatter, Facebook at Work).

This requires the development of an information system that is able to process large amounts of raw data into valuable information that is then used in decision making and which contributes to the creation of knowledge within the organisation. The data source for such a system is the reporting of individual units under accounting activities. Therefore, many studies on performance measurement mainly focus on these issues. Both traditional cash measures and non-cash measures of supply chain operations are must-have in the assessment system.

In performance measurement systems various types of metrics are used, both quantitative and qualitative. The set has evolved over the years. The impact of unpredictable business environment and the increasing pressure from consumers have led to the transition from cost-oriented strategies to customisation and value-oriented strategies. Such strategies promote other measures [16]. Supply chains have changed and the role of individual entities within supply chains has changed as well. This has increased the need for changes in performance assessment [69]. The changes also took place in performance measurement systems [16].

A well-functioning performance measurement system positively influences the quality of managerial decisions. Especially in the supply chain environment, where many decisions are strongly interdependent (trade-offs) and are dispersed throughout a pattern of business entities. At the same time—as argued by Holmberg [27]—it is impossible to carry out measurement, monitoring and evaluation of performance in the supply chain in a completely uniform manner—as if it were one single company. According to him, it would harm supply chain integration. The author proposes that the foundation for supply chain performance measurement should not be one and only system, but independent systems in which metrics are used individually but which carry out assessment in a broader system aspect, with the recognition of relationships between entities and processes. This should be supported by the use of improvement models, such as SCOR, which reveal how supply chain really works [27].

From the perspective of supply chain management it is extremely important that the performance measurement system promotes the integration of many different business areas. The implementation of an integrated performance measurement system is in practice connected with many challenges the overcoming of which is critical to the success of the process. They result from [20]:

- the process of setting targets for each measure and assessing performance against these targets;
- if the targets are consistent with each other, they may not be useful for guiding managers in their day-to-day operations;
- inconsistencies and trade-offs among the targets provided managers with conflicting signals.

The selection of metrics to measure, monitor and assess supply chain performance is determined by many characteristics. Many studies in this regard adopt the SCOR optics and point to metrics associated with the main processes in the supply chain [12, 22]. Others link metrics to specific management levels [23] and still others—to specific characteristics of the supply chain [6]. Ahmed [1] in turn indicates the phase of product life cycle and the implemented market strategy. The pragmatic approach is very important. In this approach the performance measurement system is a comprehensive strategic management tool. Consistency with strategic goals, inclusiveness, universality and measurability are the four features of an effective performance measurement system [6].

Metrics can also be linked to stakeholder groups whose interests they reflect. In these approaches there are many similar metrics that one may call basic. They are of a universal nature, such as EBITDA (Earnings Before Interest, Taxes, Depreciation and Amortisation). Their use is justified regardless of the direction of improving supply chain performance. Nevertheless, they should be supplemented with a set of specific metrics related to the direction of supply chain development including adaptive supply chain.

4 The 3V's as the Main Features of Adaptive Supply Chains

4.1 Visibility

Supply chain visibility (SCV) is a complex issue that involves people, processes, technology and information flow [75]. However, there is no single, generally accepted definition of the SCV. Some authors, e.g. Swaminathan and Tayur [61] focus their attention mainly on information flow, defining visibility as the aptitude to ensure the access to information in the scope of a supply chain. Others, still, concentrate on the features of the exchanged information, arguing that the level of the SCV is determined by the scope in which this information is accurate, trusted, timely and in a readily usable format. Zhang et al. [75] define the notion of supply chain visibility from the IT perspective as an ability to gather and analyse dispersed data as well as generate specific recommendations that refer to strategy. But undoubtedly SCV is not only dependent on IT. There are many others enablers of visibility beyond technology-based ones [4].

One of the significant aspects of the SCV is transparency with regard to resources. It means the possibility to determine their current location in the supply chain, their volume, condition as well as the readiness of their handling. Such concept of transparency refers to all resources, but it particularly concerns stocks, determining the capability of monitoring their level in any link from any other link in the supply chain [62].

Another issue connected with the SCV is the visibility in terms of demand. Demand visibility is the capability of the system to possess undistorted information on the precise demand in time, that allow partners in the supply chain to react efficiently [26].

Supply chain visibility is created at three levels. The first one concerns gathering relevant information. The second refers to finding proper communication tools in order to disclose the collected information to other enterprises. The third level involves the skill to use information for continuous improvement of the functioning of the supply chain [33].

Scholten and Schilder [59] claim that apart from information sharing, transparency also requires mutual communication and shared creation of knowledge. All these elements provide supply chains with visibility required for early detection and proper reaction to all disruptions that appear in them. The SCV is determined by the scope in which the participants of the supply chain have access to the current information and may be treated as a condition for proper reaction to changes and disruptions, thanks to which it is termed as predecessor of resistance. Moreover, managers can expect increased visibility with extensive information processing capabilities from supply chain organization's internal integration [72].

Visibility is also interpreted as detecting potential problems before they appear. It supports anticipating the potential appearance of disruptions in the supply chain [59]. Visibility also fosters anticipating delays in supplies before they happen and applying proper methods of reaction. Wilhjelm [71] states that the SCV means "see around corners". Proper visibility that involves the entire supply chain plays a vital role in its management, making it more sensitive and susceptible to control [42].

4.2 Velocity

Literature defines supply chain velocity in many ways, taking into account various aspects. The velocity of the supply chain is understood by many authors as the time that lapses from the moment of placing an order until the execution of the delivery [56]. The lead-time is thus perceived as a key indicator of the supply chain velocity [36]. With regard to the B2C relations, velocity can be interpreted as the aptitude to satisfy the needs of the final customer in a short time, and in case of business customers (B2B relations) it means delivering the terms and conditions of the contract in a short time [62]. Still, Tsironis and Matthopoulos [67] define velocity as the capability of fast execution of various processes and measures.

Velocity may be assessed on the basis of e.g. [5]:

- time of order execution,
- time devoted to each process in the supply chain,
- share of deliveries executed on time,
- stock turnover ratio.

Some authors point to two major elements of velocity: speed of reaction and speed of implementing changes in order to deliver products exactly when they are needed [25]. Juttner and Maklan [36] describe velocity as the speed with which the supply chain is able to react to events and changes taking place on the market. Scholten and Schilder [59] apply a similar definition of this notion, extending it by an element connected with time required for restoring the continuity of the operation of the supply chain after disruptions appear.

Adequate management of supply chain velocity has a great impact on its efficiency and achieving a competitive advantage [25]. Therefore, it is crucial to aim for maximising the velocity of the supply chain, also by means of:

- while selecting suppliers—concentration on their flexibility, i.e. the capability of immediate reaction, where requirements concerning various parameters of the order may change (e.g. the volume of the order) [13];
- proper selection of suppliers of key materials and services, i.e. accounting for such factors as: the distance between the supplier and the recipient's location, agreed penalties for non-provision of obligations, extra charges for accelerated deliveries, service quality standards etc. [14];
- process facilitation by means of their re-designing in order to reduce the number of operations and simultaneous execution of operations [13];
- aiming at minimising the batch volume (order volume, production batch volume or consignment volume) in order to focus on flexibility towards the economies of scale [13];
- minimising the time devoted to operations that, according to the customers, add no value [13];
- planning synchronisation in the entire supply chain [29, 31];
- establishing trust among partners in the supply chain, joint problem solving as well as facilitating quick access to information and resources necessary for proper reaction to non-standard events [34];
- replacing stocks with information to avoid potential stock shortages that denote lost opportunities, as well as excessive volume of stocks that generate unnecessary costs [14];
- data transmission in real time, which allows supply chain to limit the time required for order execution [56];
- sharing current information in the entire supply chain with the application of technologies used for electronic exchange of data, handling orders, tracking stocks, supplies etc. [29, 37].

The problem with ensuring adequate velocity in the functioning of the supply chain stems from two major reasons: its structure and priorities determined by enterprises that operate within it. The first element primarily refers to situations, where large distances between the partners' locations cause lengthened time of order execution. The second reason, on the other hand, results from the concentration of specific enterprises on their business activity, regardless of the interests of all participants in the supply chain [29].

4.3 Versatility

In the light of changing market conditions and customer requirements, supply chains must be ready to ensure flexibility and changeability of the executed operations [51]. Therefore, another feature of supply chains that is crucial in terms of acquiring and maintaining competitive advantage is versatility [46, 52]. This feature mainly involves balancing the operational capability of the supply chain with the market needs, in particular providing adequate products and services of the required quality and volume. It is also vital to adapt the offer to the individual needs of customers.

Supply chain versatility is expressed in the aptitude to cooperate with suppliers and recipients in the context of various conditions of delivery execution. This feature means the ability to maintain the operational continuity of the supply chain in particularly unfavourable conditions of the environment (e.g. high level of inflation, changes in legal provisions, unstable political situations, natural disasters etc.). Versatility also involves flexibility in the field of operational conditions. On the one hand, it is the capability of adjusting to the requirements of various suppliers, on the other—the potential to satisfy the needs of various clients [62].

Aiming for satisfying individual, frequently specific customer requirements engages maximising the variety of products offered. This is also affected by the development of technologies and the products' shortened lifecycle. Adapting the product range to the requirements of various customers by means of applying such methods as: mass individualisation, customised designs, customised configuration and postponing assembly are also connected with managing variability in the supply chain [58].

Many authors treat the notion of versatility solely as the number of various products offered to customers, but they should be understood to a greater extent. This is due to the fact that it can be executed by means of e.g. introducing diverse product features, in particular packaging, diversification in the scope of distribution channels etc. [40]. As supply chains function in the international market, the need for versatility does not solely result from the will to satisfy specific customer requirements, but also the necessity to adjust products to the legal requirements of various countries, specific climates, languages etc. [48].

5   Research Methodology

5.1   Development of the Survey Instrument

The construction of the measuring tool was initiated with drawing up a list of indicators and metrics of supply chain performance that were cited in the literature and also applied in the business practice. Next, the authors selected only those that were most frequently mentioned and that encompassed the perspective of the entire supply chain. The following step involved selecting potential indicators for each assumed dimensions of the supply chain performance, described with the use of the 3V formula, based on literature review. In effect, an initial set of statements in the questionnaire was drawn up. The list of indicators was limited on the basis of the principle "less is better" [12], according to which the system of performance measurement should be based on the minimal number of metrics and indicators.

In effect, the scale for measuring the performance of the adaptive supply chain included 23 indicators (Appendix 1). The list of indicators has been prepared based on the definition of three assumed dimensions of the supply chain performance. Questions were listed without grouping into categories. Likert's seven-level scale was used in the questionnaire to evaluate each indicator: from "strongly disagree" to "strongly agree".

5.2 Sample and Data Collection

The research was conducted using the CATI (computer assisted telephone interviews) technique. Interviews were conducted with the management staff on the basis of a structured questionnaire, the work is based on positivist paradigm. The research sample consisted of 200 Polish companies (from all Polish voivodeships) employing 50 or more employees of which 63% were medium-sized enterprises employing less than 250 employees and 37% of large enterprises employing more than 250 employees. Companies are representatives of four sectors of economy: automotive, food, furniture as well as consumer electronics and household appliances, which are among most advanced sectors in the Polish economy (leaders of Polish export). The research sample was selected in a quota random way (Table 1). The percentage of denials or unsuccessful contact attempt is 81%.

The analysis of the gathered data was conducted at two stages. Firstly, the set of indicator variables selected for measuring supply chain performance was subject to the exploratory factor analysis (EFA). The aim of EFA is to obtain a minimum number of factors that contain the maximum possible amount of information contained in the original variables used in the model, and with the greatest possible reliability [55]. The use of exploratory factor analysis allows the identification of a small number of latent variables (factors) that cannot be measured directly but are presented by observable indicators [35]. Next, based on the obtained results, a confirmatory factor analysis (CFA) was conducted, which again involved the

Table 1 Information according research sample (n = 200)

Sector of economy	Number of companies (% n)	Number of denials or unsuccessful contact attempts (% n)
Automotive	50 (25%)	373 (44%)
Food	50 (25%)	140 (16%)
Furniture	50 (25%)	152 (18%)
Consumer electronics and household appliances	50 (25%)	192 (22%)
Total	200 (100%)	857 (100%)

Source own elaboration

modification of the set of indicator variables by means of deleting those that appeared statistically insignificant or inaccurate (factor load values did not meet expectations). Data were analysed using IBM SPSS Statistic version 23.0.

6 Analyses and Results

The analysis of the gathered data was initiated with an exploratory factor analysis. The Kaiser-Meyer-Olkin (KMO) measure of sampling adequacy was 0.911, indicating a good sample size [73]. Bartlett's test of sphericity was significant ($\chi^2(253) = 1959.426$, $p < 0.000$) which indicating the variables are correlated enough for an EFA analysis—is bigger than the suggested minimum values of 0.5 [17] and 0.6 [64]. The number of factors to retain was decide using the Kaiser rule (retain factors with eigenvalues higher than 1) and scree plot analysis (Cattell's scree test). For a four-factor solution, a factor analysis was conducted with the use of a Principal Axis Factoring. The rotation of the obtained factor solution involved a method of oblique rotation Oblimin with the *Delta* parameter equalling 0. With the use of the factor loadings matrix, insignificant indicators were deleted, namely those that to no factor had a factor loading equalling the absolute value larger than 0.3 [9]: SCP14, and SCP15. Also ambiguous indicators were eliminated, namely those that had significant (though frequently relatively low) loadings for several factors: SCP16, SCP12, SCP22, SCP13, and SCP8. The final rotated factor matrix for the EFA is presented in Table 2. The use of EFA enabled the identification of four factors related to supply chain performance, namely: Responsiveness, Versatility, Visibility, and Velocity. These are three factors from the 3V formula supplemented by a brand new factor—Responsiveness.

The reliability analysis for each extracted factor (measurement scale) was made using Cronbach's alpha. In all cases Cronbach's alpha is higher than 0.60—Cronbach's alphas were 0.792, 0.718, 0.776 and 0.819 for Factors 1, 2, 3 and 4 respectively. Cronbach's alpha greater than 0.6, especially with a small number of questions, means that the set of observable variables (measured directly on Likert scale) is a reliable instrument for latent variable measurement. All the developed scales demonstrated reliabilities above the recommended threshold range of 0.6–0.7 [50].

The structure obtained in the EFA framework was verified with the use of a confirmatory factor analysis, which was aimed at evaluating a factor model that binds selected factors with constructs they are to measure. The values of the model parameters were assessed with the GLS method. In order to obtain a solution that is best suited to data, in the light of generally accepted matching criteria, the following variables with the lowest values of factor loadings (below 0.6) and with explained variances (below 0.4) were eliminated from specific factors: SCP1, SCP7, SCP6, SCP3, and SCP9.

The quality assessment of the model engaged a series of goodness-of-fit. The authors made an initial assessment of the model with the use of chi-squared statistics with reference to the number of degrees of freedom. It is often argued that

Table 2 EFA of supply chain performance

	Factor 1 Responsiveness	Factor 2 Versatility	Factor 3 Visibility	Factor 4 Velocity
SCP11	**0.860**	−0.058	-0.079	0.033
SCP3	**0.597**	0.149	0.113	−0.004
SCP10	**0.595**	−0.154	−0.160	−0.227
SCP23	**0.511**	−0.035	0.081	−0.063
SCP9	**0.421**	−0.054	0.202	−0.073
SCP1	**0.348**	−0.083	0.175	−0.001
SCP18	0.006	**−0.887**	0.109	0.042
SCP17	0.070	**−0.504**	0.000	−0.185
SCP2	0.261	0.154	**0.637**	0.013
SCP5	0.074	−0.148	**0.611**	−0.002
SCP4	0.076	−0.149	**0.595**	−0.001
SCP6	−0.060	0.070	**0.567**	−0.144
SCP7	0.017	−0.163	**0.408**	−0.060
SCP21	−0.019	0.054	0.007	**−0.960**
SCP20	0.100	−0.074	−0.019	**−0.702**
SCP19	0.021	−0.047	0.179	**−0.517**
Variance explained (%)	34.229	6.439	4.707	4.424

Bold characters indicate highly significant correlations between the factors and corresponding variables.
Source own elaboration

the model is very good when this value is smaller than 2; if it oscillates between 2 and 5—the model is considered as acceptable [18]. In the assessed model the value χ^2/df equals 1.259. The good fit of the model is also confirmed by the RMSEA equals 0.036. It is assumed that the model is acceptable if the approximation error does not exceed 0.08 [39] and good (adequate) if the value is below 0.05 [18].

Good model fit is also confirmed by such measures as CFI = 0.948, GFI = 0.959, AGFI = 0.924, which exceed the required value of 0.9 [10, 39]. Only the NFI = 0.808 reached the value below 0.9. The main drawback of the NFI is its sensitivity to the sample size (it is frequently underestimated for samples below 200) and the model's complexity (higher values are obtained for more complex models). This problem was solved by the application of the TLI, which prefers simpler models [28]. For the assessed model, the TLI exceeded the acceptance threshold and equals 0.921.

The next stage of model assessment was evaluating the theoretical validity, which involved determining the convergent validity and discriminant validity. The convergent validity is connected with the convergence of indicators measuring the same construct, and the discriminant validity helps to assess whether the indicators correlate too strongly with the measures of other constructs.

The convergent validity was evaluated on the basis of three criteria: (1) values of factor loadings (>0.7) and their significance; (2) reliability analysis: Cronbach's

Table 3 Assessment of reliability

Factor	Questionnaire statements	Standardised factor loadings	Cronbach's α	CR
Responsiveness	SCP11: The supply chain has the capacity to deliver products to the final customer exactly on time	0.740	0.729	0.801
	SCP10: The supply chain guarantees a short time from the moment of order placement to the execution of the delivery	0.831		
	SCP23: In the supply chain the level of customer satisfaction is analysed	0.696		
Versatility	SCP18: The supply chain is capable of providing products in different variants	0.807	0.718	0.750
	SCP17: The supply chain can handle non-standard orders and satisfy special customer requirements	0.742		
Visibility	SCP2: The supply chain is characterised by considerable planning accuracy	0.645	0.749	0.767
	SCP5: The supply chain can detect the appearing problem connected with order execution and deal with them	0.768		
	SCP4: In the supply chain, it is possible to track and monitor order fulfillment and related resource flow	0.753		
Velocity	SCP21: The supply chain can swiftly implement product improvements	0.850	0.819	0.856
	SCP20: The supply chain can swiftly launch a new product on the market	0.853		
	SCP19: The supply chain can quickly adapt its production capacity so as to accelerate or slow down production in its reaction to decreasing demand	0.739		

Source own elaboration

alpha and the CR reliability coefficient for specific constructs (>0.7); (3) the average variance extracted (AVE)(>0.5). Standardised factor loadings in the analysed model met the required criterion; merely two (with the SCP2 and SCP23) obtained values slightly below 0.7. All factor loadings are statistically significant (Table 3).

The results of the reliability analysis (on the basis of α-Cronbach and CR) indicate high coherence of items comprising the scales that measure four dimensions of the supply chain performance. Also the AVE was used to measure convergent reliability. Its value smaller than 0.5 means that on average there remains

Table 4 Fornell-Larcker criterion

	AVE	Responsiveness	Versatility	Visibility	Velocity
Responsiveness	0.574	(0.76)			
Versatility	0.601	0.561	(0.78)		
Visibility	0.524	0.625	0.518	(0.72)	
Velocity	0.665	0.623	0.580	0.597	(0.82)

The square root of the average variance extracted (AVE)—in brackets
Source own elaboration

more of error at the positions constituting the structure of the latent variable than the extracted variance [57]. The AVE for specific latent variables reached values from 0.524 to 0.665 (Table 4). The above results confirm that the convergent validity for all constructs is high.

The discriminant validity involved the Fornell-Larcker test, which focuses on verifying whether the AVE square root for each construct is higher than the correlations between the factors [19]. At the matrix diagonal (numbers in brackets) was filled with the AVE square root values for constructs, whereas the numbers outside the diagonal are the values of relevant correlation coefficients (Table 4). The criterion is satisfied if the number at the diagonal is highest in comparison with other numbers from own verse and column [32]. All latent variables met the described criterion.

Summarising the obtained results, it can be argued that the conditions for satisfying the model's theoretical validity are sufficient.

7 Discussion

The scale for measuring the supply chain performance that resulted from the conducted research includes four factors. Each of them portrays a different performance aspect of the adaptive supply chain, and variables that are connected with a given factor measure the level of a specific feature of a supply chain.

The first factor, called Supply Chain Responsiveness, is associated with such indicators as "The supply chain guarantees a short time from the moment of order placement to the execution of the delivery", "The supply chain has the capacity to deliver products to the final customer exactly on time", and "In the supply chain the level of customer satisfaction is analysed". This construct has no equivalent in a specific feature that complies with the 3V formula. This factor primarily refers to the aspects of the supply chain responsiveness connected with getting familiarised with customer needs as well as reaction to them (delivering products fast and in a timely manner). Lee [41] and Whitten et al. [70] wrote about creating adaptive supply chains by means of analysing the needs of both direct as well as final customers. Still, Szymczak [62] claimed that responsiveness to the needs of a customer is one of the three major directions in the evolution of supply chains (apart from flexibility and resistance to disruptions) that result from the adaptive capacity. The first factor also referred to the time of order execution as well as their timely delivery, as e.g. Basu and Wright argued [5].

The second factor, Supply Chain Versatility, includes such items as "The supply chain is capable of providing products in different variants", and "The supply chain can handle non-standard orders and satisfy special customer requirements". In its essence, this factor is close to the third element of the 3V formula—versatility. The supply chain reaching high values in the scope of this dimension is characterised by a high level of flexibility and changeability of the undertaken arrangements [51]. It also has the capacity to adapt to the requirements of various suppliers and is capable of satisfying individual, specific customer requirements [62]. This factor is also associated with the variety of the offered products that may be executed by means of e.g. implementing various product features, special packaging, diversity in terms of distribution channels, etc. [40, 58].

Another factor, Supply Chain Visibility contains such items as "The supply chain can detect the appearing problem connected with order execution and deal with them", "In the supply chain, it is possible to track and monitor order fulfillment and related resource flow", and "The supply chain is characterised by considerable planning accuracy". A supply chain reaching high values in the framework of this dimension is characterised by transparency necessary for early detection and proper reaction to all sorts of disruptions, in particular associated with order execution [59]. Ensuring visibility of all processes provides necessary information in order to make decisions and corrections in plans. This allows partners in the supply chain to identify bottlenecks, which in turn fosters immediate reaction in order to eliminate them [31]. Supply chain visibility is also connected with the ability to track the flow of resources, in particular inventories, as well as the current update of the order fulfillment status [62].

The fourth factor separated on the basis of the conducted factor analysis is Supply Chain Velocity. This construct includes such factors as "The supply chain can swiftly implement product improvements", "The supply chain can swiftly launch a new product on the market", and "The supply chain can quickly adapt its production capacity so as to accelerate or slow down production in its reaction to decreasing demand". In its content, this factor is approximate to the second element of the 3V formula—velocity. This construct is associated with the capacity of the supply chain to execute various processes and measures aimed at achieving the desired goals in a fast manner [67]. On the one hand, such velocity refers to implementing changes: the development of the currently offered products and launching new products [25]; on the other—it is associated with the ability to react to diverse events and changes on the market [36].

8 Conclusions

Adaptability as virtue enabling adjusting to all the more challenging operational conditions is one of the paramount factors that guarantee supply chains' long-term competitiveness and success. Adaptive supply chains are capable of adapting to: changes in environment, changes in the executive environment. They have the features of flexibility, responsiveness and resilience what could be considered/ reflected from a managerial perspective by the 3V formula relating to its visibility, velocity and versatility. Undoubtedly, this way of thinking can be seen as a shortcut and it really is one, but it shows in a simple way how pragmatic and utilitarian 3V formula supports management to achieve one of the most desired strategic goals for a supply chain, namely adaptability.

The goal of this paper was the elaboration of the set of metrics for supply chain performance measurement that could be used in case of adaptive supply chains. According to the conducted research the scale for measuring the supply chain performance should be expanded to four factors, namely responsiveness, versatility, velocity, and visibility (3V + R formula). Those findings support general view that adaptive supply chains are characterised by the transparency, high level of flexibility and capacity to adapt to the requirements of suppliers as well as the capacity to execute various processes and measures aimed at achieving the desired goals in a fast manner. Moreover, analyses reveal that 3 V formula need to be supplemented by yet another factor called responsiveness that relates to reaction to customer needs (delivering products fast and in a timely manner).

The scale could be used for the performance measurement of smart supply chains. It that case, however, smartness could be understood as the feature supporting supply chain adaptiveness and performance improvement. New emerging technologies, such as IoT and BigData, support information management and supply chain integration what in turn help to adjust to changing surrounding and consumer needs. Today's smart supply chain is able to capture and communicate real time data, can operate (almost) in real-time and can even anticipate customers behaviour. In the last stage smart supply chains will evolve into self-adaptive supply chains that fully rely on digital solutions and powered by artificial intelligence will be perceived as self-driving.

Acknowledgements The study was funded by the National Science Centre, Poland (grant no. 2014/13/B/HS4/03293).

Appendix 1

Questionnaire statements

Statement	Source
SCP1: The supply chain is able to limit stocks	Based on Whitten et al. [70]
SCP2: The supply chain is characterised by considerable planning accuracy	Based on Tarasewicz [65]
SCP3: The supply chain is capable of limiting wastefulness	Based on Whitten et al. [70]
SCP4: In the supply chain, it is possible to track and monitor order fulfillment and related resource flows	Own
SCP5: The supply chain can detect the appearing problem connected with order execution and deal with them	Based on Juttner and Maklan [36]
SCP6: The demand forecasts developed in the supply chain are accurate	Based on (Arif-Uz-Zaman and Ahsan [2]
SCP7: The supply chain is characterised by a large volume of mutual contacts with partners	Based on Qrunfleh and Tarafdar [54]
SCP8: The supply chain is able to foresee abrupt changes	Based on Szymczak [62]
SCP9: The supply chain can minimise total costs of delivering the product to the final customer	Based on Beamon [6]
SCP10: The supply chain guarantees a short time from the moment of order placement to the execution of the delivery	Based on Jüttner & Maklan [36]
SCP11: The supply chain has the capacity to deliver products to the final customer exactly on time	Based on Beamon [6]
SCP12: The supply chain contains a mechanism for eliminating the execution of delayed, incomplete and damaged deliveries	Based on Whitten et al. [70]
SCP13: The supply chain is capable of quick reactions and solving problems raised by the final customer	Based on Tarasewicz [65]
SCP14: The supply chain is characterised by a high level of orders that can be executed immediately from the current stocks	Based on Chae [12]
SCP15: In the supply chain receivables are swiftly paid	Based on Chae [12]
SCP16: The supply chain ensures a short reaction time in terms of customer enquiry	Based on Beamon [6]
SCP17: The supply chain can handle non-standard orders and satisfy special customer requirements	Based on Qrunfleh and Tarafdar [54]
SCP18: The supply chain is capable of providing products in different variants	Based on Qrunfleh and Tarafdar [54]
SCP19: The supply chain can quickly adapt its production capacity so as to accelerate or slow down production in its reaction to decreasing demand	Based on Qrunfleh and Tarafdar [54]

(continued)

(continued)

Statement	Source
SCP20: The supply chain can swiftly launch a new product on the market	Based on Qrunfleh and Tarafdar [54]
SCP21: The supply chain can swiftly implement product improvements	Based on Qrunfleh and Tarafdar [54]
SCP22: The supply chain offers a wide range of post-sales services	Based on Golrizgashti [21]
SCP23: In the supply chain the level of customer satisfaction is analysed	Based on Beamon [6]

References

1. Ahmed, A. M. (2002). Virtual Integrated Performance Management. *International Journal of Quality and Reliability Management, 19*(4), 414–441.
2. Arif-Uz-Zaman, K., & Ahsan, A. M. M. N. (2014). Lean supply chain performance measurement. *International Journal of Productivity and Performance Management, 63*(5), 588–612.
3. Ayers, J. B. (2004). *Supply chain project management. A structured collaborative and measurable approach*. Boca Raton: St. Lucie Press.
4. Barratt, M., & Oke, A. (2007). Antecedents of supply chain visibility in retail supply chains: A resource-based theory perspective. *Journal of Operations Management, 25*(6), 1217–1233.
5. Basu, R., & Wright, J. N. (2008). *Total supply chain management*. United Kingdom: Elsevier.
6. Beamon, B. M. (1999). Measuring supply chain performance. *International Journal of Operations & Production Management, 19*(3), 275–292.
7. Bititci, U. S., Carrie, A. S., & McDevitt, L. (1997). Integrated performance measurement systems: A development guide. *International Journal of Operations and Production Management, 17*(5), 522–534.
8. Boin, A., Kelle, P., & Whybark, D. (2010). Resilient supply chains for extreme situations: Outlining a new field of study. *International Journal of Production Economics, 126*(1), 1–6.
9. Bradley, N. (2013). *Marketing research: Tools and techniques*. Oxford: Oxford University Press.
10. Brown, T.A. (2015). *Confirmatory factor analysis for applied research* (2nd ed.). New York: The Guilford Press.
11. Butner, K. (2010). The smarter supply chain of the future. *Strategy & Leadership, 38*(1), 22–31.
12. Chae, B. (2009). Developing key performance indicators for supply chain: An industry perspective. *Supply Chain Management: An International Journal, 14*(6), 422–428.
13. Christopher, M., & Peck, H. (2004). Building the resilient supply chain. *The International Journal of Logistics Management, 15*(2), 1–14.
14. Clark, C. (2007). Getting back to basics: Top five tips for accelerating supply chain velocity. *Supply & Demand Chain Executive, 8*(4), 26.
15. D'Aveni, R. (1995). *Hyper-competitive rivalries: Competing in highly dynamic environments*. New York: FreePress.
16. De Toni, A., & Tonchia, S. (2001). Performance measurement systems. Models, characteristics and measures. *International Journal of Operations and Production Management, 21*(1–2), 46–70.
17. Field, A. (2009). *Discovering statistics using SPSS*. London: SAGE Publications.

18. Fischer, C. (2013). Trust and communication in European agri-food chains. *Supply Chain Management: An International Journal, 18*(2), 208–218.
19. Fornell, C., & Larcker, D. F. (1981). Evaluating structural equation models with unobservable variables and measurement error. *Journal of Marketing Research, 18*(1), 39–50.
20. Giovannoni, E., & Maraghini, M. P. (2013). The challenges of integrated performance measurement systems. Integrating mechanisms for integrated measures. *Accounting, Auditing and Accountability Journal, 26*(6), 978–1008.
21. Golrizgashti, S. (2014). Supply chain value creation methodology under BSC approach. *Journal of Industrial Engineering International, 10*(67), 1–15.
22. Gunasekaran, A., Patel, C., & McGaughey, R. E. (2004). A framework for supply chain performance measurement. *International Journal of Production Economics, 87*(3), 333–347.
23. Gunasekaran, A., Patel, C., & Tirtiroglu, E. (2001). Performance measures and metrics in a supply chain environment. *International Journal of Operations and Production Management, 21*(1–2), 71–87.
24. Heinemann, G. (2017). Online-Handel der Zukunft. *Der neue Online-Handel* (pp. 1–33). Wiesbaden: Springer Gabler.
25. Hines, T. (2013). *Supply chain strategies: Demand driven and customer focused.* New York: Routledge.
26. Holcomb, M. C., Ponomarov, S. Y., & Manrodt, K. B. (2011). The relationship of supply chain visibility to firm performance. *Supply Chain Forum: An International Journal, 12*(2), 32–45.
27. Holmberg, S. (2000). A systems perspective on supply chain measurements. *International Journal of Physical Distribution and Logistics Management, 30*(10), 847868.
28. Hooper, D., Coughlan, J., & Mullen, M. (2008). Structural equation modelling: guidelines for determining model fit. *Electronic Journal of Business Research Methods, 6*(1), 53–60.
29. Hudnurkar, M., & Rathod, U. (2012). Collaborative supply chain: insights from simulation. *International Journal of System Assurance Engineering and Management, 3*(2), 122–144.
30. Ivanov, D., Sokolov, B., & Kaeschel, J. (2010). A multi-structural framework for adaptive supply chain planning and operations control with structure dynamics considerations. *European Journal of Operational Research, 200*(2), 409–420.
31. Iyer, A., Seshadri, S., & Vasher, R. (2009). *Toyota supply chain management: A strategic approach to Toyota's renowned system.* New York: McGraw-Hill Education.
32. Janssen, S., Moeller, K., & Schlaefke, M. (2011). Using performance measures in innovation control. *Journal of Management Control, 22*(1), 107–128.
33. Johansson, S., & Melin, J. (2008). *Supply chain visibility. The value of information. A benchmark study of the Swedish industry.* Stockholm: KTH Royal Institute of Technology.
34. Johnson, N., Elliott, D., & Drake, P. (2013). Exploring the role of social capital in facilitating supply chain resilience. *Supply Chain Management: An International Journal, 18*(3), 324–336.
35. Jung, S. (2013). Exploratory factor analysis with small sample sizes: a comparison of three approaches. *Behavioural Processes, 97*, 90–95.
36. Jüttner, U., & Maklan, S. (2011). Supply chain resilience in the global financial crisis: An empirical study. *Supply Chain Management: An International Journal, 16*(4), 246–259.
37. Kalakota, R., Robinson, M., & Gundepudi, P. (2003). Mobile applications for adaptive supply chain: A landscape analysis. In E. Lim & K. Siau (Eds.), *Advances in mobile commerce technologies, Idea Group Inc* (pp. 298–312). USA: Hershey.
38. Kang, Y.-S., Park, I.-H., & Youm, S. (2016). Performance prediction of a MongoDB-based traceability system in smart factory supply chains. *Sensors, 16*(12).
39. Kersten, W., Blecker, T., & Meyer, M. (Eds.). (2009). *Supply chain performance management.* Berlin: Erich Schmidt Verlag.
40. Kohlberger, R., Gerschberger, M., & Engelhardt-Nowitzki, C. (2011). Variety in supply networks—Definitions and influencing parameters" In: Fang, C. H. (Ed.), *Proceedings of the 6th International Congress on Logistics and SCM systems, Kaohsiung, Taiwan* (pp. 191–203), March 07–09, 2011.

41. Lee, H. L. (2004). The triple-A supply chain. *Harvard Business Review, 82*(10), 102–112.
42. Lee, Y., & Rim, S-Ch. (2016). Quantitative model for supply chain visibility: process capability perspective. *Mathematical Problems in Engineering, 2016*, 1–11.
43. Lockamy, A., & McCormack, K. (2004). The development of a supply chain management process maturity model using the concepts of business process orientation. *Supply Chain Management: An International Journal, 9*(4), 272–278.
44. Markman, G. D., & Phan, P. H. (2011). *The competitive dynamics of entrepreneurial market entry*. Cheltenham: Edward Elgar Publishing Limited.
45. McGrath, R. M. (2013). Transient advantage. *Harvard Business Review* 64–70.
46. Momeni, E., Tavana, M., Mirzagoltabar, H., & Mirhedayatiane, S. M. (2014). A new fuzzy network slacks-based DEA model for evaluating performance of supply chains with reverse logistics. *Journal of Intelligent and Fuzzy Systems, 27*(2), 793–804.
47. Nam, T., & Pardo, T. A. (2011). Conceptualizing smart city with dimensions of technology, people, and institutions. In *Proceedings of the 12th Annual International Digital Government Research Conference on Digital Government Innovation in Challenging Times*.
48. Nielsen, N. P. H., & Holmström, J. (1995). Design for speed: A supply chain perspective on design for manufacturability. *Computer Integrated Manufacturing Systems, 8*(3), 223–228.
49. Noori, H., & Lee, W. B. (2002). Factory-on-demand and smart supply chains: The next challenge. *International Journal of Manufacturing Technology and Management, 4*, 372–383.
50. Nunnally, J. C., & Bernstein, I. H. (1994). *Psychometric theory*. New York: McGraw-Hill.
51. Nutt, B. (2004). Infrastructure resources: Forging alignments between supply and demand. *Facilities, 22*(13/14), 335–343.
52. Olugu, E. U., & Wong, K. Y. (2009). Supply chain performance evaluation: Trends and challenges. *American Journal of Engineering and Applied Sciences, 2*(1), 202–211.
53. Poirier, Ch C. (2002). Achieving supply chain connectivity. *Supply Chain Management Review, 6*(6), 16–22.
54. Qrunfleh, S., & Tarafdar, M. (2014). Supply chain information systems strategy: Impacts on supply chain performance and firm performance. *International Journal of Production Economics, 147*, 340–350.
55. Rossoni, L., Engelbert, R., & Bellegard, N. L. (2016). Normal science and its tools: Reviewing the effects of factor analysis in management. *Revista de Administração (RAUSP), 51*(2), 198–211.
56. Ruhi, U., & Turel, O. (2005). Driving visibility, velocity and versatility: The role of mobile technologies in supply chain management. *Journal of Internet Commerce, 4*(3), 95–117.
57. Ryciuk, U. (2016). *Zaufanie międzyorganizacyjne w łańcuchach dostaw w budownictwie*. Warszawa: Wydawnictwo Naukowe PWN.
58. SAP. (2002). *Adaptive supply chain networks*. http://www.carlosabrantes.com/4/upload/sapadaptivesc.pdf. Accessed January 18, 2018.
59. Scholten, K., & Schilder, S. (2015). The role of collaboration in supply chain resilience. *Supply Chain Management: An International Journal, 20*(4), 471–484.
60. Simchi-Levi, D., Kaminsky, P., & Simchi-Levi, E. (2000). *Designing and managing the supply chain. Concepts, strategies, and case studies*. Boston: McGraw-Hill/Irwin.
61. Swaminathan, J. M., & Tayur, S. R. (2003). Models for supply chains in e-business. *Management Science, 49*(10), 1387–1406.
62. Szymczak, M. (2015). *Ewolucja łańcuchów dostaw*. Poznań: Wydawnictwo Uniwersytetu Ekonomicznego w Poznaniu.
63. Szymczak, M. (2014). Smart convenience in food supply chains. In: *The International Forum on Agri-Food Logistics "Agri-Food Logistics as a Chance of Efficient Consumer Response in the Agri-Food Sector, Poznań* (pp. 178–180).
64. Tabachnick, B. G., & Fidell, L. S. (2013). *Using multivariate statistics*. Boston: Pearson.
65. Tarasewicz, R. (2014). *Jak mierzyć efektywność łańcuchów dostaw?* Szkoła Główna Handlowa – Oficyna Wydawnicza, Warszawa.
66. Tjahjono, B., Esplugues, C., Ares, E., & Pelazez, G. (2017). What does Industry 4.0 mean to supply chain? *Procedia Manufacturing, 13*, 1175–1182.

67. Tsironis, L. K., & Matthopoulos, P. P. (2015). Towards the identification of important strategic priorities of the supply chain network. *Business Process Management Journal, 21* (6), 1279–1298.
68. Upton, J. (2017). Setting sights on the smart supply chain. *Pharmaceutical Executive, 37*(3).
69. van Hoek, R. I. (1998). Measuring the unmeasurable—Measuring and improving performance in the supply chain. *Supply Chain Management: An International Journal, 3*(4), 187–192.
70. Whitten, G. D., Green, K. W., & Zelbst, P. J. (2012). Triple-A supply chain performance. *International Journal of Operations & Production Management, 32*(1), 28–48.
71. Wilhjelm, R. (2013). *Revisiting the 3Vs of supply chain: Visibility, variation and velocity.* http://www.scdigest.com/experts/ComplianceNetworks_13–10–17.php?cid=7489. Accessed January 18, 2018.
72. Williams, B. D., Roh, J., Tokar, T., & Swink, M. (2013). Leveraging supply chain visibility for responsiveness: The moderating role of internal integration. *Journal of Operations Management, 31*(7–8), 543–554.
73. Wipulanusat, W., Panuwatwanich, K., & Stewart, R. A. (2017). Exploring leadership styles for innovation: An exploratory factor analysis. *Engineering Management in Production and Services, 9*(1), 7–17.
74. Wu, L., Yue, X., Jin, A., & Yen, D. C. (2016). Smart Supply Chain Management: A review and implications for future research. *The International Journal of Logistics Management, 27,* 395–417.
75. Zhang, A. N., Goh, M., & Meng, F. (2011). Conceptual modelling for supply chain inventory visibility. *International Journal of Production Economics, 133*(2), 578–585.

Losses in Transportation—Importance and Methods of Handling

Marcin Anholcer, Tomasz Hinc and Arkadiusz Kawa

Abstract A smart supply network must be immune against the situations like losing the goods during the transportation process. Such losses may raise the necessity of increasing the number of deliveries and thus the use of fuel and pollution. The importance of this problem has been proved by the results of a quantitative research performed among Polish companies. A solution to this problem is to design a smart supply network with appropriate DSS (decision support system), based on relevant mathematical models and algorithms, that allow to reduce the number of multiple deliveries.

Keywords Perishable products · Losses in transport · Exclusions in transport DSS · Quantitative research

1 Introduction

The amounts of goods transported increase constantly. An obvious reason is the increase in production and consumption. There is, however, also another reason. If we consider the volume of transportation in some assumed period, it increases also because the delivery time is getting longer on average. It is caused mainly by the globalization. The companies localize their production far from the markets, as this allows to optimize the costs (good examples here are the producers of electronics and clothes, localizing their factories in the South-Eastern Asia, while the biggest markets are Northern America and Western Europe). Some goods are not accessible

M. Anholcer (✉) · T. Hinc · A. Kawa
Poznań University of Economics and Business,
Al. Niepodległości 10, 61-875 Poznań, Poland
e-mail: m.anholcer@ue.poznan.pl

T. Hinc
e-mail: tomasz.hinc@phd.ue.poznan.pl

A. Kawa
e-mail: arkadiusz.kawa@ue.poznan.pl

© Springer International Publishing AG, part of Springer Nature 2019
A. Kawa and A. Maryniak (eds.), *SMART Supply Network*, EcoProduction,
https://doi.org/10.1007/978-3-319-91668-2_6

in several regions (like coal, gas, oil, metals and many kinds of plants). This makes the supply networks longer, no matter if one considers geographic distances or time.

The longer are the supply networks, the higher is the risk that some product would be damaged or lose its features (and hence its value). It is an important thing to predict such situations and involve them in planning the distribution of goods. A natural way of modeling such problems is using the so-called generalized networks (see the remainder of this chapter).

There are many reasons for which the amounts of goods change during the flow through a supply network. Among others, in particular the following groups of causes may be distinguished: physical and chemical properties of transported goods, decrease in the value of transported goods corresponding with time, inappropriate transportation or storage conditions, not registered production rejects, accidents, crimes.

Physical and chemical properties of transported goods may cause the changes in their amount. One may include in this group food, in particular the agricultural products, e.g. fruits and vegetables. This is rather obvious that their quality decreases in time and it is impossible to keep them in good condition for a longer period, so the losses are inevitable. Some examples of perishable agricultural products that were analyzed by the researchers can be found in the review paper of Ahumada and Villalobos [2] (see the references therein for details). Rong, Akkerman and Grunow in [39] developed a model describing the changes in quality of fresh food. They noted that the quality is either linear or exponential function of time. Yu and Nagurney in [49] assume that the relation is exponential (see also the recent book of Nagurney, Yu, Masoumi and Nagurney [33]). Zanoni and Zavanella in [50] analyze the decision strategies for sustainable food supply chains. In particular, they ask whether it is better to freeze the food products or to keep them fresh. In particular, they give a relation between quality and time. In general, if one knows the delivery time (or at least its expected value), it is possible to predict what will be the (expected) change in the quality (or amount) of chosen product after sending it through any part of the supply network.

Next example of perishable goods are medical nuclear supplies, necessary for medical imaging. In this case, the radioactive materials are subject to decay, which causes the changes in their amount. An example of such supply network may be found in the paper of Nagurney and Nagurney [30] and in the book [33] mentioned above. In both works the authors consider the production and transportation of molybdenum-99 (^{99}Mo) and technetium-99m (^{99m}Tc).

Another perishable product is blood. The problem was described recently by Nagurney et al. in [28, 29, 33], where the authors described a bi-criteria model involving cost and risk. In fact, the blood was from the very beginning the most interesting perishable product for the researchers, as Nahmias stated in [34]. It is true that the amount of blood or blood products rather does not change over time. However, these products have a strictly defined expiration date. Some of the products, being beyond the expiration time, must be considered as impossible to use, which can be interpreted as decrease in the amount of good.

Yet another example of the products whose amount changes because of the physical and chemical features are the pharmaceuticals. This problem was studied e.g. by Masoumi, Yu and Nagurney in [27] and recently also in [33]. The authors developed a generalized network oligopoly model to analyze the Cournot competition among the drug manufacturers. Very rarely the amount of drug or other pharmaceutical product can really change, but again, as in the case of blood products, exceeding the expiration date by some amount of products is equivalent to decreasing the total volume by this amount.

Let us present another example of a perishable product. Here it is even much more visible that the amount does not change, but the percentage of the products that can be considered as having the standard value can decrease. This group of products are fast fashion apparels, studied e.g. by Nagurney and Yu in [31, 32], as well as by Choi and Chiu in [11]. Today the supply networks, in particular in the case of international companies, are considerably long. The textile industry is a good example here—in most cases the clothes are produced in the South-Eastern Asia, while the biggest markets are located in Western Europe and Northern America. This makes the time of the flow through the supply network from manufacturer to final customer to be counted in months, sometimes even more than one year. In case of fast fashion apparels it means in particular, that even if the manufacturer is able to predict what type of product will be trendy in the next season, it must be sold below the regular price, when the season ends (sales). It must be, because it is very unlikely that it will be possible to sell the same product next year. Once again—the amount of good does not change in this case, but its value decreases, so it can be treated as an equivalent situation.

Another cause of losses are the inappropriate transportation and storage conditions. Such a problem may touch each kind of product. The food and the pharmaceuticals need to be transported in a valid temperature and humidity. These conditions are often not satisfied because of the tendency to reduce the costs. Also the transportation of fragile products, like electronics or glass, requires special conditions and careful handling. Often these conditions are not satisfied—the cost reduction causes pressure on the reduction of delivery time and one of the results are often damages and losses during transportation. Not registered rejects are delivered to retailers and then often to customers. Of course these damaged products come back to producers, which can be treated as loss again—in this case the amount of transported good does not change, but the amount of sterling products is not the same as assumed when the transportation process started. The companies try to reduce the amount of rejects by introducing the quality management procedures, but it is impossible to avoid all the problems in this area.

All the above possible kinds of losses are quite easy to predict ant their distribution is usually close to uniform, which means that it is easy to predict their rate even in a short period and in the case of relatively small deliveries. Moreover, the loss rate is usually relatively small (up to few percent) if the delivery time does not exceed the standard duration. Two causes of losses described below are different—they are not that frequent and their distribution in time is usually impossible to predict. On the other hand if they occur, then usually the loss is 100%—the whole delivery must be replaced. These two reasons are accidents and crimes.

The supply network are becoming longer, in particular because of the internationalization and globalization of trade. There are several types of accidents or crimes that may occur. We will ignore the events at the production sites, as we are interested in the changes of the amount of goods occurring during the transportation.

At the beginning and at the very end of the supply network the goods are usually transported by cars. Dozens of car accidents occur almost every day in almost all the countries in the world. After many of them, the whole delivery is lost and needs to be replaced by a new one. Also a crime can be the reason of losses at this stage.

The most obvious is theft—part of delivery can be picked for example from the parking place, but quite often whole car is stolen (or even kidnapped together with the driver).

The railway accidents are so rare that they can be ignored. There are, however, crimes that can be committed, in particular in the case of bulk transportation—for example it is relatively often reported that some amount of coal was stolen from a cargo train, or that the protesting farmers scattered some agricultural products from the wagons.

In the global supply network there is also another part of the trip dangerous for the cargo—the sea transport. The products are transported between continents in the containers. And it can happen that a container falls from the ship—some estimations say about even 10,000 containers that are lost over the side of container ships, as a result of high seas, improper stowage, fire accidents and pirates (see e.g. [47]). The pirates are hijacking even whole ships. They are usually ransomed, but it usually takes much time and at least a part of the transported products loses its value.

As mentioned before—in the former cases it is difficult (or even impossible) to predict the events, and even their distribution. However, in the case of the biggest companies, these estimations become easier, as the accidents and crimes occur more often. This means that in some cases one still has a chance to predict the rate of losses and include them into the decision process.

Let us finish this section with a brief description of the results of a qualitative research performed recently in Poland. Complete results can be found in [3]. Almost all the participants of the research admitted that their companies were touched by the problem of perishable products. The list of the reasons were rather similar in each case. The most important (although not very often) problem are the thefts. Sometimes not only the products, but even the whole truck has been stolen. The owners of the transportation firms are usually insured, but clearly the insurance covers in such case only the value of the products and means of transport. The loss of customer because of delay in delivery is not included, neither are the losses in the image of the firm. Also important, an maybe most common reason of losses is the inappropriate treatment of the products during the transportation. Two types of situations occur.

A product can be damaged if not well treated, but it can be also missing because of inappropriate treatment (e.g. if it is not well secured it can fall out from a ship or truck). Another reason are accidents. Also in this case the specific insurance can be bought, but it does not cover all the types of losses. The losses resulting from delays (e.g. the losses in fresh food or medical supplies) are not often, as the transporting companies are aware that in such case very often the whole delivery is not usable.

What is interesting, such situations seem to be a more significant problem for small companies because of their lack of elasticity. It is also huge problem for the companies transporting the goods outside the European Union, in particular to the Eastern Europe countries, as the delays on the borders are very often significant in this case (sometimes equal to 2–3 days). There are also some really rare situations when two products should not be transported together, but for some reason they are. One of the examples given by the participants of the research was transportation of oil having specific smell together with flour. In such case the flour may be sometimes not usable. The respondents mentioned also the temperatures. In particular in winter, it is not always necessary to use fridges during the transport provided that the weather does not change suddenly, in particular because of the fact that the destination is the Southern Europe.

The losses of goods during the transportation process have a significant connection with the ecology. Each loss generates the increment of the traffic, as if, say, the loss ratio is r, then, in order to satisfy the demand, one needs to transport the amount of goods higher than the assumed one approximately

$$\frac{1}{1-r}$$

times. This obviously increases the number of deliveries, so also the use of energy and pollution. On the other hand, such an increase in the amount of goods also implies that the production must be bigger than necessary. It is a problem in particular in the case of food industry, where each increment of production causes the increasing need for water and other rare resources, and causes extra pollution because of the use of chemical fertilizers and pesticides.

In order to show, how important this problem is, in addition to the qualitative research mentioned above, also a quantitative research was conducted. We present its results in the following section.

2 Losses in Transportation—Results of Quantitative Research

The research was performed in Poland, in 2014, using CAWI (computer-assisted web interview) and CATI (computer-assisted telephone interview) techniques. Respondents were 300 managers responsible for storage in logistics companies operating in Poland, selected randomly. Questionnaire was preceded by a pilot in which 7 representatives from researchers and experts in the field of logistics have been chosen. In the given year, Eurostat reports show approximately 140k companies in the transport and warehousing sector, after excluding pipeline and passenger transport this number goes down to 94k. Sample size for the project was set on a 383 companies, to achieve confidence level and significance on 0.95 and 5% respectively.

As we can see in Table 1, close to 30% respondents checked the last two boxes ("important" and "very important"), from which car accidents, burglary and thefts were noted as the most painful reasons for losses. From the other hand, bottom end touches maritime and railroad transport.

We can conclude from Table 2, that according to respondents, the highest percentage of losses is generated, again, by exogenous factors such as accidents, thefts and burglary, but is clear that endogenous ones, like inappropriate transport conditions and trucks overloading, however not selected in the previous table as most important reasons, appear dangerously often.

Table 3 is less prone for cognitive biases than Tables 1 and 2. When asked to mark the importance of losses, managers answers probably reflect their emotional

Table 1 Significance of loss factors

	N/A	Completely not important	Not important	Modest	Important	Very important
Accidents	4	73	33	45	31	104
Car accidents	79	9	32	35	32	103
Car burglary	10	93	30	34	31	92
Car theft	10	137	19	24	13	87
Goods stolen by employee	9	125	28	22	20	86
Client refused to accept delivery because of the delay	10	85	41	39	31	84
Car overloading: too heavy load	9	82	38	47	31	83
Inappropriate conditions: careless treatment of sensitive goods	104	28	13	28	39	78
Inappropriate transport conditions	8	94	23	45	46	74
Goods lost	11	129	35	22	24	69
Car overloading: too large goods	12	121	33	38	19	67
Unspotted losses (before transport begins)	4	87	36	58	40	65
Inappropriate conditions: temperature	102	47	16	32	36	57
Co-transportation of certain goods	13	134	31	29	27	56
Goods physico-chemical features	8	150	35	26	21	50
Value lost due to expiring	8	178	22	25	18	39
Inappropriate conditions: humidity	102	61	17	47	25	38
Railroad accidents	82	170	5	10	4	19
Ship hijacking	11	252	2	4	3	18
Sea accidents	81	176	4	10	6	13
Train burglary	11	252	5	9	2	11
Others	259	24	0	1	1	5

Table 2 Frequency of loss factors

	N/A	Never	Rarely	Sometimes	Often	Very often
Inappropriate transport conditions	5	181	44	32	16	12
Car accidents	171	1	70	19	17	12
Inappropriate conditions: careless treatment of sensitive goods	183	22	31	31	14	9
Car burglary	5	191	62	18	6	8
Car overloading: too heavy load	9	154	66	41	12	8
Car theft	5	255	16	4	3	7
Car overloading: too large goods	5	206	41	22	9	7
Accidents	5	166	74	18	20	7
Inappropriate conditions: temperature	180	41	25	27	11	6
Unspotted losses (before transport begins)	4	157	75	36	12	6
Client refused to accept delivery because of the delay	4	185	62	25	10	4
Goods stolen by employee	5	232	39	8	2	4
Goods physicochemical features	5	214	43	17	7	4
Goods lost	6	242	32	6	1	3
Railroad accidents	173	107	3	2	2	3
Inappropriate conditions: humidity	181	52	26	19	9	3
Co-transportation of certain goods	6	224	45	11	1	3
Train burglary	9	268	8	3	0	2
Value lost due to expiring	5	222	34	19	8	2
Sea accidents	174	111	3	1	0	1
Others	251	32	1	3	2	1
Ship hijacking	10	276	4	0	0	0

reaction on a given event; road accidents and thefts are psychologically heavier than customers complaints on not sufficient load treatment, but as we take a look on the losses ratio to all transportation, "unspotted losses" rises to top 5, from financial perspective, as well as losses due to car overloading.

Table 4 presents losses ratio in company losses as a whole, not only those related to transportation. Here, distribution is dominated by car derived ones and as in Table 3, "unspotted" category.

In the described study, most common methods for handling the risk of losses were reducing and avoiding the risk, indicated by about 7 and 8% of respondents respectively. Risk transfer was the method of choice for those situations that could be coped with insurance.

We also built contingency tables and checked independence of various factors using standard χ^2 test at the significance level of 0.05. It came out that in most cases the factors are pairwise dependent, which means that the problem of losses in

Table 3 Rate of losses in transport relatively to the overall value of delivered goods

	N/A	0–1%	1–5%	5–10%	10–20%	20%+
Unspotted losses (before transport begins)	169	69	38	10	2	2
Goods physicochemical features	224	38	18	5	3	2
Others	286	0	2	2	0	0
Client refused to accept delivery because of the delay	196	50	32	3	4	5
Car theft	262	24	1	1	1	1
Goods stolen by employee	242	34	10	2	2	0
Inappropriate transport conditions	193	38	47	8	3	1
Inappropriate conditions: careless treatment of fragile goods	211	33	36	6	4	0
Inappropriate conditions: temperature	227	26	32	3	2	0
Inappropriate conditions: humidity	240	35	11	3	1	0
Ship hijacking	286	2	1	1	0	0
Car overloading: too heavy load	173	60	35	12	5	5
Car overloading: too large goods	218	42	22	3	1	4
Co-transportation of certain goods	237	40	12	1	0	0
Car burglary	205	59	17	6	2	1
Train burglary	278	8	3	0	1	0
Accidents	180	57	37	6	2	8
Railroad accidents	282	6	1	1	0	0
Sea accidents	286	1	2	0	0	1
Car accidents	181	56	37	7	2	7
Goods lost	253	27	6	3	1	0
Value lost due to expiring	232	39	14	1	2	2

Table 4 Rate of losses in transport relatively to overall losses

	N/A	0–5%	5–10%	10–25%	25–50%	50%+
Unspotted losses (before transport begins)	172	85	24	7	2	0
Goods physicochemical features	228	49	7	3	2	1
Others	286	1	2	1	0	0
Client refused to accept delivery because of the delay	197	59	25	7	0	2
Car theft	265	23	1	0	1	0
Goods stolen by employee	244	40	4	2	0	0
Inappropriate transport conditions	196	48	40	5	1	0

(continued)

Table 4 (continued)

	N/A	0–5%	5–10%	10–25%	25–50%	50%+
Inappropriate conditions: careless treatment of fragile goods	216	47	20	4	3	0
Inappropriate conditions: temperature	230	29	28	2	1	0
Inappropriate conditions: humidity	242	37	9	1	1	0
Ship hijacking	286	3	1	0	0	0
Car overloading: too heavy load	173	87	21	6	1	2
Car overloading: too large goods	221	51	12	2	3	1
Co-transportation of certain goods	237	45	8	0	0	0
Car burglary	210	57	15	5	1	2
Train burglary	279	9	1	0	1	0
Accidents	187	66	26	4	3	4
Railroad accidents	284	4	1	1	0	0
Sea accidents	287	2	1	0	0	0
Car accidents	188	64	26	5	3	4
Goods lost	257	29	3	1	0	0
Value lost due to expiring	236	42	11	1	0	0

transport should be treated as a whole, no matter what causes were reported by the companies.

3 Solution to the Problem

As we can see, the losses in transportation are an important issue, which was confirmed by the research performed among Polish companies. On the other hand, we know that those losses are not only a financial problem, but may also have an influence on the ecology. The question is, how to solve this problem, i.e. how to design supply networks so that one can reduce the side effects to minimum.

One of the ways to deal with this problem is to design reliable supply networks, in which the future losses are somehow assumed. If one sends an amount of goods a little greater than the requested amount, it is possible to satisfy the demand without sending another transport.

Of course, it is very simple if one considers one source and one destination. However, if the supply network has a more complicated structure, with many sources and destination and with some transshipment points, it is necessary to plan the deliveries very carefully.

The mathematical models that are used in this case are based on the concept of generalized networks. In a standard setting, one assumes that the amount of goods leaving the source is exactly the same as the amount reaching the destination. In such situation one can use the following mathematical model, in which $V(G)$

denotes the set of nodes of the network G (i.e., the sources, destinations and transshipment points), $E(G)$—the links between them (i.e., arcs of the network), x_{ij} is the amount of flow, c_{ij}—the unit transportation cost, u_{ij}—maximum capacity of connection between nodes and b_i represents the supply if positive or demand if negative (cf. [1, 4]):

$$f(x) = \min \sum_{ij \in E(G)} c_{ij} x_{ij},$$

s.t.

$$\sum_{ij \in E(G)} x_{ij} - \sum_{ji \in E(G)} x_{ji} = b_i, i \in V(G),$$

$$0 \le x_{ij} \le u_{ij}, ij \in E(G)$$

Of course, this is a simplest model and one can consider other possible settings, where the demand is treated as random, the costs are nonlinear or there are more than one objectives (see e.g. [4] for various variants). If one assumes losses during transportation, it is necessary to introduce the so-called multipliers r_{ij}, representing the part of commodity that reaches the destination (it is defined independently for every connection). In such situation, the model takes the form:

$$f(x) = \min \sum_{ij \in E(G)} c_{ij} x_{ij},$$

s.t.

$$\sum_{ij \in E(G)} x_{ij} - \sum_{ji \in E(G)} r_{ji} x_{ji} = b_i, \quad i \in V(G),$$

$$0 \le x_{ij} \le u_{ij}, ij \in E(G)$$

It looks very similar to the previous one, but this slight modification in the form implies quite significant algorithmic issues (in particular, when one considers more complex networks and more complicated function relations between the variables and parameters). While in the former case the optimal solutions lead to the transportation networks having the form of forests (i.e. acyclic graphs, see Fig. 1), in the latter case they take form of augmented forests, where every component has one extra arc and so it is no longer a tree, because it contains a cycle (Fig. 2). Again—this looks like a small modification, but causes significant algorithmic issues.

The various approaches to model and solve such problems (also in more general settings, not only in the case of simple linear relations and also in less obvious applications than supply networks) were presented in various articles.

Fig. 1 Sample network and
its spanning forest *Source*
Own work based on [4]

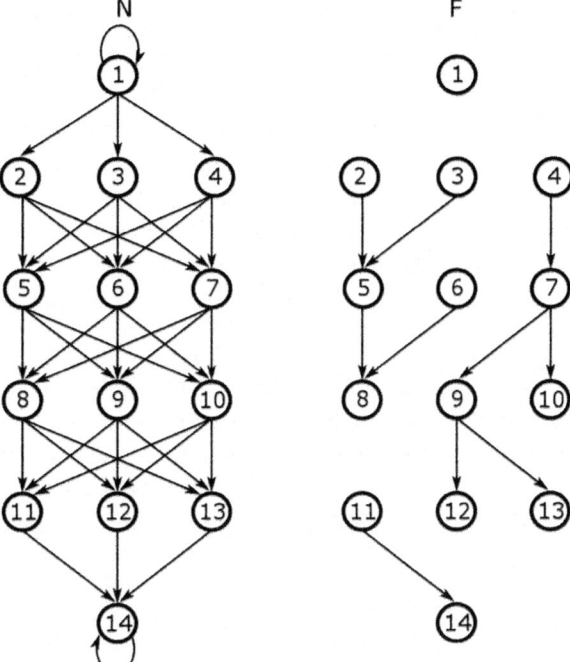

Fig. 2 Sample network and
its spanning A–forest *Source*
Own work based on [4]

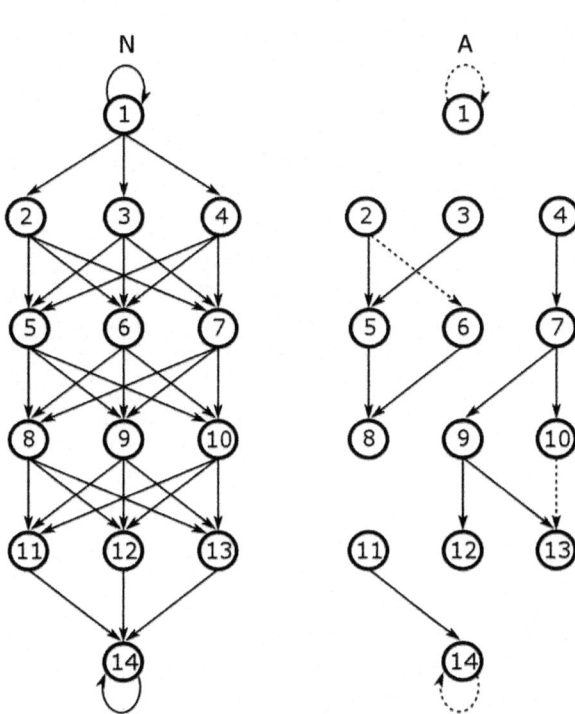

Probably the first paper where a solution method for Generalized Minimum Cost Flow Problem (GMCFP) was published was the one of Jewell [22], where a primal-dual algorithm was presented. Several authors studied various variants and aspects of Generalized Network Simplex Method (see e.g. [1, pp. 593–590, 8, pp. 494–497, 14, 15, 23, 24]). Bertsekas and Tseng in [10] proposed a relaxation method. Goldfarb and Lin in [19] presented a combinatorial interior point method. In 2002, Wayne published the paper [48], where he introduced the first combinatorial, polynomial time algorithm for the general variant of GMCFP. The algorithm repeatedly cancels minimum ratio circuits, where the sum of costs is placed in the numerator and the sum of the reciprocals of the residual capacities is the denominator of the ratio. This is a generalization of the previously known minimum ratio cycle-cancelling algorithms, see [48] for details.

A special case of GMCFP is the Generalized Circulation Problem (GCP). In this case one skips the costs (or assume that all the unit costs are equal). Cohen and Megiddo in [12] presented an algorithm that solves the GCP by repeatedly solving instances of the so-called Uncapacitated Generalized Transshipment Problem (UGTP). The algorithm can be also used to obtain an approximation and in this case it is strongly polynomial if used a constant factor approximation. First polynomial algorithms for GCP were presented by Goldberg et al. in [18]. Glover et al. in [17] characterized the properties special generalized network problems, in which a dual feasible basic solution could be determined in one 'pass' through the network. This approach allowed to use the dual method and poly-ω technique to pure network problems. Same techniques were later applied also to a special class of the generalized network problems (see below).

A special case of GMCFP is the Generalized Transportation Problem (GTP), where the underlying network is bipartite, one of the partition sets consists only from the supply nodes and another from demand nodes. The transshipment nodes are not involved in this type of problems. First complex description of the problem was published by Dantzig. In [13] (pp. 413–432) he presented the problem, analyzed its properties and proposed a solution method. What is worth mentioning, he considered a more general setting of the problem, called Weighted Distribution Problem (WDP), where the multipliers are involved on the both sides (supply and demand). This modification, however, does not increase significantly the computational effort, as in order to transform of this problem to the GTP form it is enough to linearly transform the original variables. The same version of the problem (i.e., the WDP) was also considered by Lourie in [26] and in several papers of Sikora (see below). In early publications (see e.g. [7, 26]) the parametric method was used to solve the GTP (the parameter was used to compute potentials, as well as the changes in the flow). Balas in [6] published a dual algorithm for the GTP. He specialized the dual method and the poly-ω technique for this kind of problems. One of the elements presented in the paper was a simple technique for solving the parametric version of the problem. Some computational aspects of solving the GTP were considered by Glover and Klingman in [16]. In particular, the authors proposed new procedures of pricing out and changing the basis, allowing to avoid the parametrization, the most popular technique. This issues were further discussed by

Balachandran and Thompson in [5]. Among others, these authors proposed a simplification allowing to pass the cycle only once (while it was necessary to pass it three times in the original algorithm). Gottlieb in [20] proposed a solution method for GTP where ordinary TP are solved repeatedly. Sikora in [41] considered the WDP and proposed an algorithm, called generating path algorithm. The same author further developed this method in [42] for the problems with income criterion and slightly different structure of constraints. Next generalization of the method (improvement tree algorithm) were presented in [43] (problem with bounded flows), and [44] (two-stage GTP).

Also other kinds of generalized flow problems attracted some attention of researchers. Thompson and Sethi in [46] considered the constrained GTP, where additional linear constraints appear. They used the pivot and probe algorithm to solve this kind of problem. Gupta in [21] analyzed the same type of problem with additional bounds on variables. He presented a solution method that allowed to store all necessary information in the form of ordinary GTP and one extra $p \times p$ matrix, instead of storing the whole base matrix inverse. The generalized version of Maximum Flow Problem were considered recently by Restrepo and Williamson in [38], where the authors proposed a simple GAP-canceling algorithm.

Also the algorithms used to solve the ordinary flow problems may be sometimes applied, without introducing many substantial changes, to the generalized flow problems. For instance, Nielsen and Zenios in [35] presented massively parallel solutions of the linear network problems, comparing three independent algorithmic schemes. Goldfarb and Jin [19] proposed a new scaling algorithm for MCFP, which can be applied after some small modifications also to the GMCFP.

The nonlinear versions of GTP have not attracted too much attention until the recent years. Zenios and Censor in [52] studied the TP and GTP with separable, nonlinear transportation costs and provided massively parallel row-action algorithms (similar topics were considered by Zenios in [51]). Rowse in [40] considers a version of GTP with demand and supply functions defined, respectively, for each destination and each supply point. The survey of nonlinear continuous allocation problems, including transportation type problems with quadratic costs, can be found in the paper of Patriksson [36]. In particular, the main problem considered in this paper has the form of nonlinear GTP with quadratic objective function and single sourcing. This kind of problem arises also as a subproblem while solving the nonlinear GTP with decomposition techniques, such as Lagrangian relaxation, or Benders decomposition [9]. Sikora in [45] analyzed the problem of plant production. The model considered in this paper has form of the Weighted Distribution Problem where a separable concave objective function is maximized, where the nonlinear components are functions in the total amounts leaving suppliers and delivered to the destinations. Such problem is obviously transformable to the form of nonlinear GTP with separable convex costs. Qi [37] presented the so-called A-forest method for nonlinear, convex GTP. Recently several algorithms for various kinds of generalized flow problems were presented by Anholcer [4].

The appropriate mathematical models and algorithms should be the elements of the company's decision support system (DSS), aiding the management to properly

Fig. 3 Scheme of a DSS.
Source Own work based on
[25]

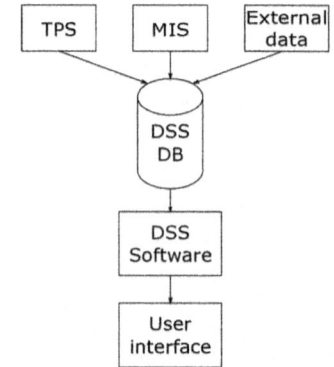

design and manage the supply network. DSS in turn are the elements of the firm's information system (see Fig. 3). To be more specific, they are parts of the appropriate software.

According to Laudon and Laudon [25], there are three main categories of information systems: systems designed for the strategic level, systems designed for the management level and systems designed for the operational level. The systems designed for the strategic level are supposed to help the top management to make right strategic decisions. They are used in particular to analyze long-term global trends, not necessarily connected directly with the organization's present performance. The systems designed for the management level are used by the middle-level managers for controlling and decision making. These systems rely on periodic reports and help to make good middle-term decisions. Sometimes they support the routine, repetitive activities, sometimes they are used in the exceptional, unique situations. The systems designed for the operational level are used to help the low-level managers in the usual, quotidian activities. These may be elementary transactions like paying salaries, sales, invoicing or flow of the materials in the warehouse during the completion of an order. The main purpose of the systems on this level is to keep all the routine activities of the organization in order. DSS are one of the systems used on the management level.

A typical DSS consists of four layers. First layer is the data collection system. Next layer is the DSS database, being a collection of historical and present data. Third layer is the DSS software system, containing all the applications necessary for transforming the data into useful reports, using the appropriate models and algorithms to solve the decision problems (in the case discussed in this paper, these are the problems consisting in planning the deliveries so that one can avoid sending multiple transports). The last layer is the user interface. The scheme of a typical DSS has been presented in Fig. 3.

4 Conclusion and Future Work

The losses in transportation, among other those caused by an inappropriate combining of goods that should not be transported together, have a significant influence on ecology. They raise the need for transporting more goods in order to satisfy the demand. This could cause, in turn, the increase in number of deliveries, the increment of fuel use and finally—the increasing pollution.

The importance of the problem is not only a theoretical issue. It was proved by a quantitative research performed among Polish companies that the problem is present in practice. In the case of 28% companies the problem can be treated as significant (they indicated it as very important or important).

A possible solution to the problem is to design of a smart supply network, connected with an appropriate decision support system, involving in particular appropriate mathematical models and efficient algorithms.

There are at least two possible directions of a future research in this area. One could perform similar quantitative research among the companies in other countries to recognize what is the significance of the problem and how the companies deal with it. It is also possible to design new mathematical models and algorithms that allow to efficiently plan the deliveries. Finally, one could work on designing new, efficient DSS to logistic companies.

Acknowledgements The paper was written with financial support from the National Center of Science (Narodowe Centrum Nauki)—the grant no. DEC-2014/13/B/HS4/01552.

References

1. Ahuja, R. K., Magnanti, T. L., & Orlin, J. B. (1993). *Network flows. Theory, algorithms and applications*. New Jersey: Prentice Hall, Englewood Cliffs.
2. Ahumada, O., & Villalobos, J. R. (2009). Application of planning models in the agri-food supply chain: a review. *European Journal of Operational Research, 196*(1), 1–20.
3. Anholcer, M., & Kawa, A. (2017). Identyfikacja strat i braków powstających w transporcie - wyniki badań przeprowadzonych metodą jakościową [Identification of losses and faults in transport: results from qualitative research]. *Gospodarka Materiałowa i Logistyka, 12*(2017), 22–29.
4. Anholcer, M. (2017). *Optymalizacja przewozów produktów szybko tracących wartość – modele i algorytmy [Optimization of transport of perishable products – models and algorithms]*. Poznań: Wydawnictwo Uniwersytetu Ekonomicznego w Poznaniu.
5. Balachandran, V., & Thompson, G.L. (1974). A note on Computational simplifications in solving generalized transportation problems, by Glover and Klingman. Discussion Papers 66, Northwestern University, Center for Mathematical Studies in Economics and Management Science, http://www.kellogg.northwestern.edu/research/math/papers/66.pdf. Cited 28 July 2015.
6. Balas, E. (1966). The dual method for the generalized transportation problem. *Management Science Series A (Sciences), 12*(7), 555–568.
7. Balas, E., & Ivanescu, P. L. (1964). On the generalized transportation problem. *Management Science Series A (Sciences), 11*(1), 188–202.

8. Bazaraa, M. S., Jarvis, J. J., & Sherali, H. D. (2010). *Linear programming and network flows* (4th ed.). Hoboken, New Jersey: Wiley.
9. Benders, J. F. (1962). Partitioning procedures for solving mixed-variables programming problems. *Numerische Mathematik, 4*, 238–252.
10. Bertsekas, D. P., & Tseng, P. (1988). Relaxation methods for minimum cost ordinary and generalized network flow problems. *Operations Research, 36*(1), 93–114.
11. Choi, T.-M., & Chiu, C.-H. (2012). Mean-downside-risk and mean-variance newsvendor models: implications for sustainable fashion retailing. *International Journal of Production Economics, 135*(2), 552–560.
12. Cohen, E., & Megiddo, N. (1994). New algorithms for generalized network flows. *Mathematical Programming, 64*, 325–336.
13. Dangalchev, C. A. (1996). Partially-linear transportation problems. *European Journal of Operational Research, 91*, 623–633.
14. Elam, J., Glover, F., & Klingman, D. (1979). A strongly convergent primal simplex algorithm for generalized networks. *Mathematics of Operations Research, 4*(1), 39–59.
15. Glover, F., Karney, D., Klingman, D., & Napier, A. (1974). A computation study on start procedures, basis change criteria, and solution algorithms for transportation problems. *Management Science, 20*(5), 793–813.
16. Glover, F., & Klingman, D. (1973). A note on computational simplifications in solving generalized transportation problems. *Transportation Science, 7*(4), 351–361.
17. Glover, F., Klingman, D., & Napier, A. (1972). Basic dual feasible solutions for a class of generalized networks. *Operations Research, 20*(1), 126–136.
18. Goldberg, A.V., Plotkin, S.A., Tardos, E. (1988). Combinatorial algorithms for the generalized circulation problem. In *SFCS'88 Proceedings of the 29th Annual Symposium on Foundations of Computer Science* (pp. 432–443).
19. Goldfarb, D., & Lin, Y. (2002). Combinatorial interior point methods for generalized network flow problems. *Mathematical Programming Series A, 93*, 227–246.
20. Gottlieb, E. S. (2002). Solving generalized transportation problems via pure transportation problems. *Naval Research Logistics, 49*(7), 666–685.
21. Gupta, R. (1978). Solving the generalized transportation problem with constraints. *Zeitschrift für angewandte Mathematik und Mechanik, 58*(10), 451–458.
22. Jewell, W. S. (1962). Optimal flow through networks with gains. *Operations Research, 10*(4), 476–499.
23. Johnson, E. L. (1966). Networks and basic solutions. *Operations Research, 14*(4), 619–623.
24. Langley, R.W. (1973). Continuous and integer generalized network flow problems. Ph.D. thesis, Georgia Institute of Technology.
25. Laudon, K. C., & Laudon, J. P. (2006). *Management information systems. Managing the digital firm* (9th ed.). Upper Saddle River, NJ: Pearson-Prentice Hall.
26. Lourie, J. R. (1964). Topology and computation of the generalized transportation problem. *Management Science Series A (Sciences), 11*(1), 177–187.
27. Masoumi, A. H., Yu, M., & Nagurney, A. (2012). A supply chain generalized network oligopoly model for pharmaceuticals under brand differentiation and perishability. *Transportation Research E, 48*(4), 762–780.
28. Nagurney, A., & Masoumi, A. H. (2012). Supply chain network design of a sustainable blood banking system. In T. Boone, V. Jayaraman, & R. Ganeshan (Eds.), *Sustainable supply chains: Models, methods and public policy implications* (pp. 49–72). London: Springer.
29. Nagurney, A., Masoumi, A. H., & Yu, M. (2012). Supply chain network operations management of a blood banking system with cost and risk minimization. *Computational Management Science, 9*(2), 205–231.
30. Nagurney, A., & Nagurney, L. S. (2012). Medical nuclear supply chain design: A tractable network model and computational approach. *International Journal of Production Economics, 140*(2), 865–874.

31. Nagurney, A., & Yu, M. (2011). Fashion supply chain management through cost and time minimization from a network perspective. In T.-M. Choi (Ed.), *Fashion supply chain management: Industry and business analysis* (pp. 1–20). Hershey, Pennsylvania: IGI Global.
32. Nagurney, A., & Yu, M. (2012). Sustainable fashion supply chain management under oligopolistic competition and brand differentiation. *International Journal of Production Economics, 135*(2), 532–540.
33. Nagurney, A., Yu, M., Masoumi, A. H., & Nagurney, L. (2013). *Networks against time. Supply chain analytics for perishable products.* New York: Springer.
34. Nahmias, S. (1982). Perishable inventory theory: A review. *Operations Research, 30*(4), 680–708.
35. Nielsen, S. S., & Zenios, S. A. (1993). Proximal minimizations with D-functions and the massively parallel solution of linear network programs. *Computational Optimization and Applications, 1,* 375–398.
36. Patriksson, M. (2008). A survey on the continuous nonlinear resource allocation problem. *European Journal of Operational Research, 185,* 1–46.
37. Qi, L. (1987). The A-Forest Iteration Method for the stochastic generalized transportation problem. *Matematics of Operations Research, 12*(1), 1–21.
38. Restrepo, M., & Williamson, D. P. (2009). A simple GAP-canceling algorithm for the generalized maximum flow problem. *Mathematical Programming Series A, 118,* 47–74.
39. Rong, A., Akkerman, R., & Grunow, M. (2011). An optimization approach for managing fresh food quality throughout the supply chain. *International Journal of Production Economics, 131*(1), 421–429.
40. Rowse, J. (1981). Solving the generalized transportation problem. *Regional Science and Urban Economics, 11,* 57–68.
41. Sikora, W. (2003). Algorytm generujących ścieżek dla zagadnienia rozdziału [Generating paths algorithm for generalized transportation problem]. In A. Całczyński (Ed.), *Metody i zastosowania badań operacyjnych '2002, Prace naukowe Politechniki Radomskiej* (pp. 283–304).
42. Sikora, W. (2004). Algorytm indeksowy dla zagadnienia rozdziału z kryterium dochodu [Index algorithm for generalized transportation problem with income criterion]. In E Panek (Ed.), Matematyka w Ekonomii. *Zeszyty Naukowe Akademii Ekonomicznej w Poznaniu* (Vol. 41, pp. 361–375).
43. Sikora, W. (2008). Algorytm drzewa poprawy dla zagadnienia rozdziału z ograniczoną pojemnością, pól [Improvement tree algorithm for capacitated generalized transportation problem]. In W. Sikora (Ed.), Z prac Katedry Badań Operacyjnych. *Zeszyty Naukowe Akademii Ekonomicznej w Poznaniu* (Vol. 104, pp. 130–146).
44. Sikora, W. (2010). Dwuetapowe zagadnienie rozdziału z kryterium dochodu [Two-stage generalized transportation problem with income criterion]. In W. Sikora (Ed.), Z prac Katedry Badań Operacyjnych, Zeszyty Naukowe Akademii Ekonomicznej w Poznaniu (Vol. 138, pp. 60–76).
45. Sikora, W. (2012). Optymalizacja produkcji roślinnej jako nieliniowe zagadnienie rozdziału [Optimization of plant production as a nonlinear generalized transportation problem]. *Metody Ilościowe w Badaniach Ekonomicznych, XII*(1), 184–193.
46. Thompson, G. L., & Sethi, A. P. (1986). Solution of constrained generalized transportation problems using the pivot and probe algorithm. *Computers & Operations Research, 13*(1), 1–9.
47. Waters, D. (2007). *Supply chain risk management. Vulnerability and resilience in logistics.* London, Philadelphia: Kogan Page Limited.
48. Wayne, K. D. (2002). A polynomial combinatorial algorithm for generalized minimum cost flow. *Mathematics of Operations Research, 27*(3), 445–459.
49. Yu, M., & Nagurney, A. (2013). Competitive food supply chain networks with application to fresh produce. *European Journal of Operational Research, 224*(2), 273–282.
50. Zanoni, S., & Zavanella, L. (2012). Chilled or frozen? Decision strategies for sustainable food supply chains. *International Journal of Production Economics, 140*(2), 731–736.

51. Zenios, S. A. (1994). Data parallel computing for network-structured optimization problems. *Computational Optimization and Applications, 3,* 199–242.
52. Zenios, S. A., & Censor, Y. (1991). Massively parallel row-action algorithms for some nonlinear transportation problems. *SIAM Journal on Optimization, 1*(3), 373–400.

Part III
Green Supply Networks

The Importance of Intra-firm Relationships in Green Supply Chain Management—A Conceptual Framework

Tomasz Surmacz and Bogdan Wierzbiński

Abstract The study focuses on crucial aspects related to factors having direct influence on the willingness of entities to become integrated in green supply chains and also to factors hindering cooperation in this respect. The modern supply chains must be smarter which means interconnected and taking better decisions on the basis of collected data. The idea of functioning within a smart network consists in an enterprise taking advantage of its own resources in order to maintain its development in the long run and is related to pro-ecological trends in the global society. The presented model analyses relational factors having influence on the performance of green supply chains. The study also points out the role of knowledge management in raising employees' awareness and the relation between trust and commitment.

Keywords Green supply chain management · Collaboration · Trust
Commitment

1 Introduction

The development of companies and conditions related to globalising economies, but also ecological awareness of societies forced business organisations to undertake a course of action which, as a consequence, leads to the creation of green supply chains. Entities operating within this type of chain, pursuing their vested interests are subject to pressure of external conditions, but also internal ones. Research shows that organisational culture and the awareness of the management and employees constitutes a catalyst for such changes. A common feature of enterprises functioning in green network structures is the inclusion of issues related

T. Surmacz (✉) · B. Wierzbiński
Department of Marketing and Entrepreneurship, University of Rzeszów,
Rzeszów, Poland
e-mail: toms@ur.edu.pl

B. Wierzbiński
e-mail: bowie@ur.edu.pl

© Springer International Publishing AG, part of Springer Nature 2019   131
A. Kawa and A. Maryniak (eds.), *SMART Supply Network*, EcoProduction,
https://doi.org/10.1007/978-3-319-91668-2_7

to ecology in inter-organisational activities. The main goal of this paper is to build a GSCM implementation model that will form the basis for conducting future empirical research.

2 Resource-Based Theory and Network Theory

Commenced in the 1950s, the discussion about resource determinants of development [58] also known as Resource Based Theory, has been widely characterised and scientifically scrutinised since the 1980s [1, 3, 4, 72, 77, 80, 86]. In that period the subject of the discussion was the issue of interdependence between the structure of resources of a business organisation and its competitive advantage. As a result, an enterprise stopped being perceived as a set of resources and competences (distinguishing it from competitors), which enabled the creation of competitive advantage and the capability to compete on a dynamic market. Within the resource based approach to the process of creating advantage, the resources of an organisation are understood as valuable, rare, difficult for emulation and, pertinently, having a significant contribution to the improvement of the situation of an enterprise in a competitive space, simultaneously being a considerable source of value of an enterprise and its growth [7, 18, 43, 44].

The concept of dynamic capabilities of smart companies seems to be quite significant in the resource based approach from the perspective of creating inter-organisational relations. The concept consists in the existence of two types of capabilities: operational and dynamic [30]. Operational capabilities allow companies to function on a daily basis as they refer to their capabilities to coordinate the day-to-day business operations being the basis of its activity, i.e. operational planning, basic logistical processes. Cepeda and Vera claim that these are routine processes, constituting a lower level of the company's capabilities with regard to dynamic capabilities [10].

Resource based theories stress the significance of any resources owned by an organisation or at its disposal. A special role in conducting economic activity is played by human resources (intellectual capital within an organisation) and intangible resources (e.g. reputation, trust, organisational culture, intellectual property rights, technology, know-how). The specialist literature on the subject suggests that the process of creating competitive advantage is based, first of all, on knowledge and competences of the employees who are able to create the necessary intangible resources [16].

The process of knowledge creation cannot be limited to a single enterprise due to e.g. high costs and inefficiency of this type of approach. If an enterprise wants to take an actual part in the processes of increasing value within a chain and to be one of its strong links, it is forced to apply a smart approach. Functioning within a structure defined in such a way is conditional upon creating and sharing knowledge.

The term "network" is an ambiguous concept and, seemingly, there are many characteristics of this phenomenon. One of them is simply a finite set of elements connected with each other (or partially connected) by a set of lines, which is how a relation is formed (limited by the connections of points in a specific network) [49]. An example of an organisation formed based on a logistical network is a supply chain.

Relations formed in network systems can be crucial in the process of internationalisation of entities [23]. The alternative for the independence of an enterprise is the creation of inter-organisational relations which can be a source of competitive advantage, while the key factor in the development of a network structure appears to be the identity and reputation of an organisation [63]. The understanding of this process in the context of globalising economies significantly increased the importance of such structures, while the development and diffusion of information technologies facilitated communication between network structure participants [65]. It has to be mentioned that the much desired synergy effect can occur between resources themselves, but also between resources and capabilities [15].

The characterised resources allow an enterprise to effectively produce and offer value for specific markets or segments. Advantage in this context comes from two sources: first of all, from distinctive competences (creating the competitive potential of an enterprise), which are interrelated with capabilities and other resources [17]. It is possible to characterise numerous categories of resources participating in the process of delivering value onto the market, however, in the contemporary economic organisation it is people with capabilities, knowledge and competences who play a very significant role. Another relevant factor is also the culture of an organisation or its policy, tools of control and the relations established with consumers and suppliers.

3 Integrative Relations Between Organisations in the Context of Forming Long-Lasting Relations Within a Supply Chain

According to the specialist literature on the subject, it may be stated that cooperation between entities operating on the market is one of the most relevant concepts related with the development strategy and survival of enterprises in a volatile environment and also the ongoing globalisation processes participated by the company. The created relations between entities form relational and social capital, contributing to the improvement of intangible resources in an enterprise, which leads to relationships based on cooperation in the context of increasing competitiveness and innovativeness of entities and, consequently, leads to the improvement of the performance of whole sectors [69]. It is possible to discern two characteristic ways of influencing the behaviour of a partnership relation: first of all, spillover effects and the establishment of proper relations between entities [22]. The specialist literature on the subject shows rivalry as a feature which underlies free market economy with the more and more frequently accompanying cooperation of enterprises. However, it cannot replace market competition, but is of complementary nature, being an example of understanding business through relations and cooperation in the management process [22]. When considering business networks, three characteristic approaches are usually taken into account [79]:

- structural competitive advantage,
- characteristic of cohorts of objects,
- characteristic of the general structure of many relations.

The most relevant feature in network analysis is the focus on relations formed between the units constituting its structure, the developed patterns and implications caused by the existing relations. Thanks to such an approach, it is possible to have a wider perspective on the whole structure formed as a result of the creation of the relations [78]. At this point, it is important to highlight features related to social changes which constitute the information-technology paradigm of network [9]. Given that the network paradigm is based on cooperation, each partner should contribute certain value in the created relation, a value needed by others which does not occur within the enterprise. In this way enterprises attempt to capitalise on advantages which result from cooperation, however, at the same time they can compete in certain areas choosing convergent confrontational goals, consequently leading to competitive processes. The instability of created relations makes the formed structures undergo numerous reconfigurations over time. The specialist literature suggests that an alternative for the independence of an enterprise is the creation of inter-organisational bonds which can be the source of competitive advantage. It has to be mentioned that the synergy effect can occur between resources themselves, but also between resources and capabilities [15]. The cooperation of enterprise allows them to achieve collaborative advantages [27, 36, 81, 82]. Proper approach to cooperation brings considerable benefits. The most important advantages are: first of all, increased efficiency and effectiveness of an organisation, access to resources, knowledge and organisational capabilities of other partners within the network provided for the entity involved in the relation. Simultaneously, it is vital to point out the significance of the learning process within a network and the increase of the spectrum of strategic alternatives and innovations (product-related, organisational, process-related, marketing) in the context of the process of resource coordination. Special attention should be paid also to the increase in the potential of involved entities to acquire always new ways of competitive advantage [41].

The paradigm of network focuses on the existence of numerous business eco-systems, which interact in the process of competition, sharing, to some extent, resources with others, implementing at the same time their own strategic goals. By applying network relations with other market participants as a developmental concept, it is possible to attain greater openness, which can have both positive and negative consequences, especially when the position of an organisation within a structure and the character of relations between network participants are analysed [15].

The way in which bonds are formed can be characterised as a closed cycle of negotiations—obligations, involvement and implementation. All the elements in the cycle undergo evaluation at every stage [61]. The created relations and their architecture condition the development of enterprises and influence their competitiveness [38, 83]. There seem to be two types of relations: relationships and interactions [42]. Relations between network participants have a direct impact on the quality of implemented processes within a structure, while the implemented concept of management boosts effectiveness in the scope of process coordination among other network participants. An essential issue here is control mechanisms, trust within strategic alliances and the phenomenon of common learning of the network structure [47]. The participation of an enterprise in a smart network based on dynamic capabilities is an intentional choice resulting from the awareness of the competitive potential of a company and the willingness to multiply profits on the basis of a well-though-out strategy and the competitive potential of a chain. In this case, the environment is stabilised, thanks to which both enterprises applying to an already existing structure and the chain can create successfully operating entities, however, the chain shows a tendency to acquire the missing or insufficient resources as a result of the incorporation of a new entity. The enterprise also seems to be interested in the resources at the chain's disposal, which leads to a win-win situation, in which the catalyst for the closeness is the difference in the developmental potential and the awareness of advantages resulting from joint activities.

Another approach to the involvement with network participants stems from the negative influence of the environment on a unit and the lack of possibilities resulting from the utilisation of the internally existing competitive potential in the developmental process of a company. In this case, the environment of an enterprise forces it to behave resourcefully, which is caused by the lack of developmental opportunities with regard to external threats and identified weaknesses of an organisation.

Enterprises can also enhance their capabilities by participation in strategic alliances, which results in the improvement of managerial skills, access to strategic information and participation in business networks [5]. Effective reaction to market needs is possible thanks to the coordination of activities between the parties, creation of relations with clients and implementation of logistics processes [70].

An increasingly tighter integration into supply chains occurs when parties realise that everyone can draw benefits from the relations in a sustainable way in the long run and develop the partnership within wide and open communication. Cooperation within a network takes place when two or more entities are co-responsible for planning, managing and directing their actions. Such companies should cooperate in the scope of planning and implementing tasks to create an integrated supply chain and resources so that they can function in collaboration [59]. Hertz [32] understands integration as a process of coordinating actions and resources so that organisations can function in cooperation. Kahn and Mentzer [37] see integration simply as combining parts into one coherent whole. Integration is a process of collaboration, optimising all activities (internal and external) related to the provision of higher value to the end customer. Such an approach is based on the premise that no company can be competitive if its suppliers and recipients within the chain are not competitive either as no company seems to be isolated from others. The results of the company's activity depends, to a large extent, on its ability to react quickly and properly to challenges and opportunities, which are offered by the competition, but also by its suppliers and customers. It has to be mentioned that the customer is the main driving force behind integrative processes in supply chains.

Obviously, companies have to be willing to cooperate, although it is not enough to ensure integration since sharing their own resources and joint investments are also required. Efficient integration calls for mutual understanding, joint vision, sharing resources and achievement of not only individual, but also collective goals [67]. Integration cannot exist without cooperation. Simatupang and Sridharan [66] consider cooperation in supply chains in terms of five perspectives: process improvement, information sharing, incentive alignment, decision synchronization and integrated supply chain processes. Cao and Zhang [8] offer seven indicators of cooperation in supply chains: information sharing, goal congruence, decision synchronization, incentive alignment, resource sharing, collaborative communication and joint knowledge creation. The last indicator constitutes the basis for building innovation within supply chains, thereby creating competitive advantage. It has to be remembered that, even if possible, not always a higher level of integration of activities is desirable. Mouritsen et al. [51] claim that it may not always influence results and supply chain management in a positive way as a lot depends on the environments in which the chains are operating.

4 Supply Chain Management Versus Green Supply Chain Management

In an attempt to define a supply chain, the following might serve as a proper description: "…network of organizations that are involved, through upstream and downstream linkages, in the different processes and activities that produce value in the form of products and services in the hands of the ultimate consumer" [13]. Managing a supply chain, in turn, according to the classic definition by Christopher [12] seems to be the management of relationships with suppliers, recipients and customers in order to provide the highest value for the customer at lowest costs for the whole supply chain.

Supply chain management (SCM) includes all functions and processes which allow delivering a product or a service to the market. It seems to be a matter of debate how to define the term SCM precisely and whether it is a technique, strategy or a philosophy of operation, but, from the point of view of this study, it seems to be most important that the relational approach is applied in the explanation of supply chain networking capabilities. SCM has evolved for many years with new functions and tasks beings inscribed in the traditionally understood approach. One of the most noticeable examples is the coinage of the term Green Supply Chain Management (GSCM) as a way to include ecological issues into the topic of supply chains.

Andic et al. [2] think that GSCM means "minimizing and preferably eliminating the negative effects of the supply chain on the environment". However, Zhu et al. [90] consider GSCM to be "an important new archetype for enterprises to achieve profit and market share objectives by lowering their environmental risks and impacts while raising their ecological efficiency".

GSCM is described most often as the direct involvement of companies together with suppliers and customers in joint planning in order to decrease the negative impact on the environment in logistical and production processes. Activities of this type entail tighter cooperation in order to decrease impact on the environment related to flows in the operational activities of a company. To put it simply, GSCM means allowing for activities for the sake of the environment in the integrative practices within supply chains. Apart from the term "GSCM", there are many other in common use, all of them referring to a similar scope of activities, however, sometimes emphasising different areas. Their common feature is the inclusion of issues related to ecology into inter-organisational activities. Some examples may include: sustainable supply network management; supply chain environmental management; environmental purchasing; green logistics; sustainable supply chains.

Before the implementation of GSCM, companies have to e.g. consider what their starting point should be, the level of ecological awareness and the interdependencies and interactions between partners in the chain. It has to be emphasized that in contrast to environmental management understood in the traditional way, the concept of green supply chains includes all the stages of a product life cycle: from mining a raw material, the design, production, the distribution stage, consumers' use of the product, to the ultimate disposal after the end of a product life cycle. The specialist literature on the subject explains the basic differences in the scope of operations and processes within traditional and green supply chain management (Table 1).

There are numerous motives for undertaking actions aiming at making supply chains "green". Some of them are of internal nature, others result from the

Table 1 Differences between the green supply chain management and conventional supply chain management

Characteristics	Green SCM	Conventional SCM
Objectives	Ecological and economic	Economic
Ecological optimization	High ecological impacts	Integrated approach low ecological impacts
Supplier selection criteria	Ecological aspects long term relationship	Price short term relationship
Cost pressure	High	Low
Flexibility	Low	High
Speed	Low	High

Source Luthra et al. [45]

environment of companies. Business entities believe that GSCM can help to improve their results; in some cases, these could be actions undertaken merely to improve a company's image [14]. Literature on the subject presents the following indicators: legislation and legislative requirements [34, 91], pressure from customers [28, 56], increase in the level of integration with suppliers [73], improvement of reputation [6] and cost cutting [21, 62].

Apart from factors facilitating the decision about creating green supply chains, there are also limiting factors. Walters distinguishes two types of risk in supply chains: internal and external. The external ones refer to economic, social, environmental or political issues, while the internal ones are related to customers and suppliers. Most often identified barriers for Green Supply Chain Management are: initial investments [53, 87], low-quality organisational culture [33], lack of support for GSCM initiatives from the management [92], lack of knowledge about green practices [33, 35], fear of failure [60], lack of support from governmental organisations [39], problems with acquiring appropriate suppliers and low involvement on the suppliers' side [52] and disbelief in environmental benefits [76]. Some authors suggest that motivators are more important for enterprises than barriers when it comes to decisions about GSCM [46].

5 The Significance of Relations Between Partners in the Creation of Green Supply Chains—Conceptual Framework

The main motive to take up the topic of this study were the results of a previous research on factors determining the success of supply chains. The research was conducted in 2016 using CAWI method (Computer Assisted Web Interview). The sample (271 enterprises) was drawn from enterprises belonging to the so-called smart specializations of the Podkarpackie province with strategic importance for the development of the region (IT industry and Automotive industry). The questions prepared on the basis of literature studies have been presented on the Likert scale (7-point), representing a series of equivalent positions being indicators of the measured, one-dimensional hidden variable. One of the analysed variables was trust. It was pointed out as one of the most important factors in building successful supply chains. For IT industry the mean was 5.86 (n = 137, Std. Dev. = 0.78). And the mean for automotive industry was 5.71 (n = 134, Std. Dev. = 0.86). According to the graphs (Fig. 1), enterprises belonging to the IT industry seemed to be more aware of the importance of trust as a company's strategic resource in the management process and this can suggest they take trust more seriously. In the group of micro-enterprises employing up to 9 people the average answer is the lowest which can be understood as the least important factor for company's success in creating supply chains. This can be explained by low levels of organizational culture and

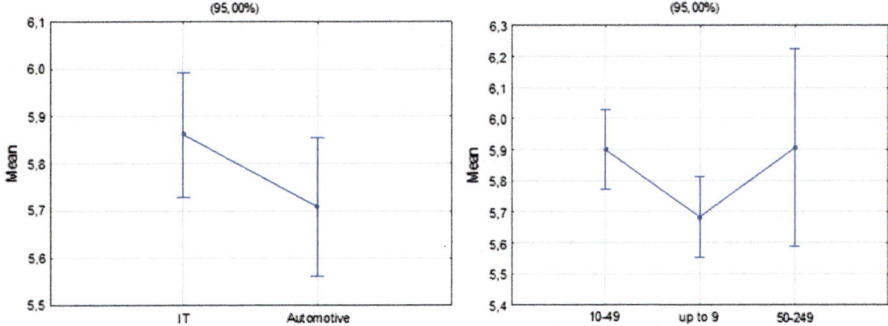

Fig. 1 Trust as a strategic resource that allows building relationships in the network in the opinion of the surveyed enterprises belonging to smart specializations in the Podkarpackie province. *Source* Own calculations

lack of experience in long term relationships. With the increase of enterprises' size the trust factor was gaining on importance.

The results of the presented research (especially the significance of trust) were the impulse for further investigations, this time taking into consideration "greening" of supply chains. Having analysed the specialist literature on the subject, the authors of the present study created a model (conceptual framework) presenting the influence of key variables on the performance of green supply chains. The theoretical model discussed below (Fig. 1) will be empirically verified during further research.

The achievement of success by a company cannot result only from its capabilities or resources. No company is devoid of suppliers and recipients, so it is crucial to understand the significance of the relation with other companies in order to build competitive advantage [75]. Client-supplier relations are essential for green supply chain management. Supporting strong relations based on cooperation between the purchaser and suppliers is of utmost importance for improving the performance of the supply chain. Relational capital allows better communication in supply chains, decreases surveillance costs and strengthens cooperation. Relational resources are connected with creating knowledge. Partnership relations, mutual trust and commitment can influence the creation of knowledge and innovative solutions in a supply chain in the scope of green solutions. And vice versa— increased level of knowledge can positively influence employees' beliefs and, as a consequence, the increase in commitment and, indirectly, in trust. Hence the first component of the model, i.e. SUPPLY CHAIN KNOWLEDGE MANAGEMENT (Fig. 2).

Strong ties between partners can lead to the development of new knowledge about how to solve problems related to green processes [74], which can cause an increase in motivation and faith in the purposefulness of their actions. Partners frequently do not know whether the undertaken actions actually make sense. If there are not available data concerning the results of conducted studies, there will be

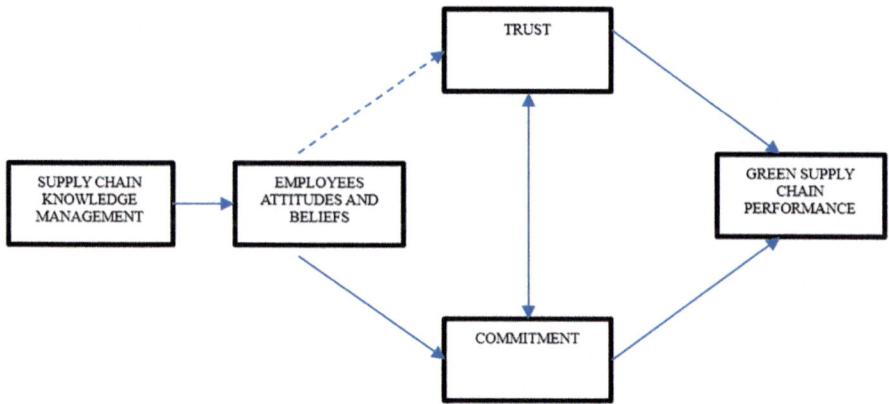

Fig. 2 Conceptual framework. *Source* Own elaboration

no positive stimuli for further work. Sharing knowledge can be a factor which induces an increase in commitment. When collaborating, partners have to establish good relations voluntarily and refrain from opportunistic behaviour at the cost of others.

Seemingly, there is a paucity of research on the nature of interpersonal relations between supply chain participants and their impact on the performance of these chains. Most studies deal with company-company relations, which is noticed by Mentzer [48], who states that it is necessary to do more research on cooperation in supply chains from the interpersonal perspective in order to understand better the conditions of interaction between employees. Human factor, in turn, is usually analysed solely with reference to top-management. It seems to be a certain limitation since the implementation of solution often lies in the hands of rank-and-file employees and it is their attitude, commitment and faith which can play a crucial role. Cooperation culture is based more significantly on good will or social norms than on formal agreements or organisational norms [89]. Interpersonal relations according to Wish et al. [84] are based on the following indicators: power symmetry/asymmetry, the degree of cooperation and friendliness maintained within the relationships, intensity of the relationships, the domain in which the relationship has been established (social or professional perspective).

In order to obtain advantages from sharing knowledge between organisation, it is vital for all the involved parties to establish close cooperation. Numerous studies recognise benefits obtained from bilateral learning between partners, which can lead to innovation and added value [54, 71]. The key issue here is to ensure that the management can appreciate the significance of green issues [31]. Nevertheless, also employees responsible for the implementation of assigned tasks have to understand them and to see the purposefulness of their actions. Employees within an organisation should understand and accept the implementation of GSCM since significant factors might also include attitudes, expectations, personalities and previous experiences of those who influence mutual relations and are able to shape them.

Additionally, Knight [40] claims that those factors combined with knowledge have impact on the ability to cooperate. Therefore, in the offered model EMPLOYEES' ATTITUDES AND BELIEFS are considered to be a key factor, conducive to developing relations based on TRUST and COMMITMENT.

Relations between companies constitute a specific type of intangible network resources [25]. Su et al. [68] state that the quality of relations in a supply chain can be understood as the extent to which parties involved in a relation can commit themselves to create a long-term relation which can function smoothly. Many authors indicate trust and cooperation as key components of the quality of relation in a supply chain [19, 55]. Others add factors such as satisfaction [26] or atmosphere [85]. Most frequently it is trust and commitment that is indicated as the distinguishing features of the strength of relations between enterprises.

Trust is necessary for intensified collaboration and information exchange. Hausman and Johnston [29] define trust as confidence in the integrity and reliability of another party, rather than confidence in the partner's ability to perform a specification. Young-Ybarra and Wiersema [88] claim that trust is based on three components: dependability, predictability and faith. According to Grayson et al. [24], trust is simply the faith that the partner is benevolent and honest. Other authors also point to benevolence, but instead of honesty, they focus on credibility, which refers to the faith in the capabilities of a partnership to act in order to achieve the results of planned actions in an efficient way [57].

Special attention should also be paid to threats related to trust. Gargiulo and Ertug [20] indicate three of them: relational inertia, resource misallocation and malfeasance. The willingness of maintaining good relations can result in ignoring signals foreshadowing problems, ceasing to monitor the performance and being "lulled into a sense of security"). Therefore, relational relationships (regardless of a wide range of positive aspects) can also have negative facets.

There seems to be a strong relation between trust and commitment. Morgan and Hunt [50] claim that commitment is "an exchange partner believing that an ongoing relationship with another is so important as to warrant maximum efforts at maintaining it". Trust and commitment are connected by a mutual relation [11]. Trust is a factor which triggers commitment [64]. Morgan and Hunt [50] think that the simultaneous presence of trust and commitment influences effectiveness and efficiency. In the presented model, there is a close correlation between trust and commitment.

6 Conclusions

The activity of stakeholders and market changes often enforce the creation of network structures, allowing for ecological issues, but also the creation of strategies defining the direction of growth, taking into consideration contemporary concepts of sustainable development and social responsibility of the business sector. The presented conditions of creation of green supply chains within the offered model will be empirically verified. It seems that research conducted so far pays insufficient attention to issues of interpersonal relations in network structures. Lack of knowledge about the need for certain actions and their effectiveness may lead to reluctance in the scope of their implementation, to the resulting decrease in commitment and, as a consequence, to lower effectiveness of supply chains. Taking these factors into account may help companies to modify their management systems with the knowledge about supply chains and to improve their communication. To achieve better decisions taking in smart supply chains more attention must be paid to partners focused on knowledge as well as trust and commitment building mechanisms.

References

1. Acedo, F. J., Barroso, C., & Galan, J. L. (2006). The resource-based theory: Dissemination and main trends. *Strategic Management Journal, 27*(7), 621.
2. Andic, E., Yurt, O., & Baltacıoglu, T. (2012). Green supply chains: Efforts and potential applications for the Turkish market. *Resources, Conservation and Recycling, 58*, 50–68.
3. Barney, J. (1991). Firm resources and sustained competitive advantage. *Journal of Management, 17*(1), 99.
4. Barney, J. B. (2001). Resource-based theories of competitive advantage: A ten-year retrospective on the resource-based view. *Journal of Management, 27*(6), 643–650.
5. Betz, F., Khalil, T., Hosni, Y., & Mostafa, H. E. (2010). *Creating and managing a technology economy* (pp. 361–362). World Scientific Publishing Co Pte Ltd.
6. Bhool, R., & Narwal, M. S. (2013). An analysis of drivers affecting the implementation of green supply chain management for the Indian manufacturing industries. *International Journal of Research in Engineering and Technology, 2*(11), 242–254.
7. Borchert, O. (2008). Resource-based theory: Creating and sustaining competitive advantage. *Journal of Marketing Management, 24*(9–10), 1041–1044.
8. Cao, M., & Zhang, Q. (2011). Supply chain collaboration: Impact on collaborative advantage and firm performance. *Journal of Operations Management, 29*(3), 163–180.
9. Castells, M. (2010). *The rise of the network society* (pp. 69–70). Wiley-Blackwell.
10. Cepeda, G., & Vera, D. (2007). Dynamic capabilities and operational capabilities: A knowledge management perspective. *Journal of Business Research, 60*, 426–437.
11. Chen, J. V., Wang, C. L., & Yen, D. C. (2013). A causal model for supply chain partner's commitment. *Production Planning & Control, 25*, 800–813.
12. Christopher, M. (1998). *Logistics and supply chain management: Strategies for reducing costs and improving service.* Financial Times Prentice Hall.
13. Christopher, M. (2005). *Logistics and supply chain management, creating value-adding networks* (p. 17). Harlow: Financial Times Prentice Hall.
14. Cyran, K., & Dybka, S. (2015). Społeczna odpowiedzialność biznesu jako czynnik ograniczający rozwój przedsiębiorstwa (Corporate Social Responsibility as a factor restricting company's development). *Zeszyty Naukowe SGGW w Warszawie. Polityki Europejskie, Finanse i Marketing, 14*(63), 7–18.
15. Czakon, W. (2007). *Dynamika więzi międzyorganizacyjnych przedsiębiorstwa (The dynamics of intra-organizational links of an enterprise)* (p. 27). Katowice: Wydawnictwo Akademii Ekonomicznej im. Karola Adamieckiego.
16. Czerniachowicz, B. (2012). Budowanie konkurencyjności przedsiębiorstwa w podejściu zasobowym na podstawie firmy „a" (The development of enterprise's competitiveness by resource-based view-quoting the example of enterprise A). *Studia i Prace Wydziału Nauk Ekonomicznych i Zarządzania. Wydawnictwo Naukowe Uniwersytetu Szczecińskiego, 25*, 287–302.

17. Day, G. S., & Wensley, R. (1988). Assessing advantage: A framework for diagnosing competitive superiority. *Journal of Marketing, 52*(2), 1–20.
18. Furman, J. L., Porter, M. E., & Stern, S. (2002). The determinants of national innovative capacity. *Research Policy, 31*(6), 899–933.
19. Fynes, B., de Burca, S., & Marshall, D. (2004). Environmental uncertainty, supply chain relationship quality and performance. *Journal of Purchasing & Supply Management, 10,* 179–190.
20. Gargiulo, M., & Ertug, G. (2006). The dark side of trust. In R. Bachmann & A. Zaheer (Eds.), *Handbook of trust research.* London: Edward Elgar.
21. Gonzalez-Benito, J., & Gonzalez-Benito, O. (2005). Environmental proactivity and business performance: An empirical analysis. *Omega International Journal of Management Science, 33*(1), 1–15.
22. Gorynia, M., & Jankowska, B. (2007). Wpływ klasterów na konkurencyjność i internacjonalizację przedsiębiorstw (The influence of business clusters on the competitiveness and internationalizations of enterprises). *Gospodarka Narodowa, 7–8,* 1–8.
23. Gorynia, M., & Jankowska, B. (2008). *Klastry a międzynarodowa konkurencyjność i internacjonalizacja przedsiębiorstwa (Business clusters versus international competitiveness and internationalization if an enterprise).* Difin: Warszawa.
24. Grayson, K., Johnson, D., & Chen, D. F. (2008). Is firm trust essential in a trusted environment? How trust in the business context influences customers. *Journal of Marketing Research, 45*(April), 241–256.
25. Gulati, R. (1999). Network location and learning: The influence of network resources and firm capabilities on alliance formation. *Strategic Management Journal, 20*(5), 397–420.
26. Gummesson, E. (1987). The new marketing—Developing long-term interactive relationships. *Long Range Planning, 20*(4), 10–20.
27. Gunn, A., & Mintrom, M. (2013). Global university alliances and the creation of collaborative advantage. *Journal of Higher Education Policy & Management, 35*(2), 179–192.
28. Hajikhani, M., Wahat, N. W. B. A., & Idris, K. B. (2012). Considering on green supply chain management drivers as a strategic organisational development approach; A Malaysian perspective. *Australian Journal of Basic and Applied Sciences, 6*(8), 146–165.
29. Hausman, A., & Johnston, W. J. (2010). The impact of coercive and non-coercive forms of influence on trust, commitment, and compliance in supply chains. *Industrial Marketing Management, 39,* 519–526.
30. Helfat, C. E., & Peteraf, M. A. (2003). The dynamic resource-based view: Capability lifecycles. *Strategic Management Journal, 24*(10), 997–1010.
31. Henneberg, S. C., Naudé, P., & Mouzas, S. (2010). Sense-making and management in business networks—Some observations, considerations, and a research agenda. *Industrial Marketing Management, 39*(3), 355–360.
32. Hertz, S. (2001). Dynamics of alliances in highly integrated supply chain networks. *International Journal of Logistics: Research and Applications, 4*(2), 239.
33. Hsu, C. W., & Hu, A. H. (2008). Green supply chain management in the electronic industry. *International Journal of Science and Technology, 5*(2), 205–216.
34. Jain, V. K., & Sharma, S. (2014). Drivers affecting the green supply chain management adaptation: A review. *Journal of Operations Management, 13*(1), 54–63.
35. Jayant, A., & Azhar, M. (2014). Analysis of the barriers for implementing green supply chain management (GSCM) practices: An interpretive structural modelling (ISM) approach. *Procedia Engineering, 97*(1), 2157–2166.
36. Jones, G. J., Edwards, M., Bocarro, J. N., Bunds, K. S., & Smith, J. W. (2017). Collaborative advantages: The role of interorganizational partnerships for youth sport nonprofit organizations. *Journal of Sport Management, 31*(2), 148–160.
37. Kahn, K. B., & Mentzer, J. T. (1996). Logistics and interdepartmental integration. *International Journal of Physical Distribution & Logistics Management, 26*(8), 9.
38. Kawa, A., & Matusiak, M. (2016). Analiza relacji sieciowych w organizacji opartej na wiedzy (Network Relationships Analysis in a Knowledge-Based Organization). *Problemy Zarządzania, 14*(4(64)), 98–119.

39. Khidir, A. T., & Zailani, S. (2009). Going green in supply chain towards environmental sustainability. *Global Journal of Environmental Research, 3*(3), 246–251.
40. Knight, L. A. (2000). Learning to collaborate: A study of individual and organizational learning, and interorganizational relationships. *Journal of Strategic Marketing, 8,* 121–138.
41. Kordha E., & Elmazi L. (2009). Information and communication technologies as an incentive for improving relationships in business to business markets. *China-USA Business Review, 8* (2) (Serial No. 68), 9–10.
42. Krzyżanowski, L. J. (1999). O podstawach kierowania organizacjami inaczej: Paradygmaty, modele, metafory, filozofia, metodologia, dylematy, trendy (About the fundamentals of managing organizations in a different manner: Paradigms, models, metaphors, philosophy, methodology, dilemmas, trends) (pp. 165–166). Warszawa: Wydawnictwo Naukowe PWN.
43. Lev, B. (2002). Intangibles at a crossroads: What's next? *Financial Executive, 18*(2), 35–39.
44. Lev, B., Radhakrishnan, S., & Zhang, W. (2009). Organization capital. *Abacus, 45*(3), 275–298.
45. Luthra, S., Kumar, V., Kumar, S., & Haleem, A. (2011). Barriers to implement green supply chain management in automobile industry using interpretive structural modeling technique: An Indian perspective. *Journal of Industrial Engineering and Management, 4*(2), 231–257.
46. Maryniak, A. (2017). *Zarządzanie zielonym łańcuchem dostaw (Green Supply Chain Management).* Poznań: Wydawnictwo Uniwersytetu Ekonomicznego w Poznaniu.
47. Mellat-Parast, M., & Digman, L. A. (2008). Learning: The interface of quality management and strategic alliances. *International Journal of Production Economics, 114*(2), 820–829.
48. Mentzer, J. T., Foggin, J. H., & Golicic, S. L. (2000). Collaboration: The enablers, impediments, and benefits. *Supply Chain Management Review, 4*(4), 52–58.
49. Mitchell, J. C. (1969). The concept and use of social networks. In *Social networks in urban situations* (pp. 1–50). Manchester: University of Manchester Press.
50. Morgan, R. M., & Hunt, S. D. (1994). The commitment-trust theory of relationship marketing. *Journal of Marketing, 58,* 20–38.
51. Mouritsen, J., Skjøtt-Larsen, T., & Kotzab, H. (2003). Exploring the contours of supply chain management. *Integrated Manufacturing Systems, 14*(8), 686–695.
52. Mudgal, R. K., Shankar, R., Talib, P., & Raj, T. (2010). Modeling the barriers of green supply chain practices: An Indian perspective. *International Journal of Logistics Systems and Management, 7*(1), 81–107.
53. Muduli, K., Govindan, K., Barve, A., & Geng, Y. (2013). Barriers to green supply chain management in Indian mining industries: A graph theoretic approach. *Journal of Cleaner Production, 47,* 335–344.
54. Nasr, E. S., Kilgour, M. D., & Noori, H. (2015). Strategizing niceness in co-opetition: The case of knowledge exchange in supply chain innovation projects. *European Journal of Operational Research, 244*(3), 845–854.
55. Naude, P., & Buttle, F. (2000). Assessing relationship quality. *Industrial Marketing Management, 29*(4), 351–361.
56. New, S., Green, K., & Morton, B. (2002). An analysis of private versus public sector responses to environmental challenges of the supply chain. *Journal of Public Procurement, 2*(1), 93–105.
57. Nyaga, G. N., Whipple, J. M., & Lynch, D. F. (2010). Examining supply chain relationships: Do buyer and supplier perspectives on collaborative relationships differ? *Journal of Operations Management, 28*(2), 101–114.
58. Penrose, E. T. (1959). *The theory of the growth of the firm.* New York: Wiley.
59. Pramatari, K. (2007). Collaborative supply chain practices and evolving technological approaches. *Supply Chain Management: An International Journal, 12*(3), 210.
60. Rao, P., & Holt, D. (2005). Do green supply chains lead to competitiveness and economic performance? *International Journal of Operations and Production Management, 25*(9), 898–916.
61. Ring, P. S., & Van De Ven, A. H. (1994). Developmental processes of cooperative interorganizational relationships. *Academy of Management Review, 19*(1), 90–118.
62. Routroy, S. (2009). Antecedents and drivers of green supply chain management implementation in a manufacturing environment. *Journal of Supply Chain Management, 6*(1), 20–35.

63. Rzemieniak, M. (2016). Tożsamość w sieciach organizacyjnych (The identity of organizational networks). *Zeszyty Naukowe, Organizacja i Zarządzanie, Politechnika Śląska, 90,* 125–142.
64. Saleh, M. A., Ali, M. Y., & Mavondo, F. T. (2014). Drivers of importer trust and commitment: Evidence from a developing country. *Journal of Business Research, 67*(12), 2523–2530.
65. Scott, N., Baggio, R., & Cooper, C. (2008). *Network analysis and tourism: from theory to practice.* Clevedon, Buffalo, NY: Channel View Publications.
66. Simatupang, T. M., & Sridharan, R. (2005). An integrative framework for supply chain collaboration. *International Journal of Logistics Management, 16,* 257–274.
67. Stank, T. P., Keller, S. B., & Daugherty, P. J. (2001). Supply chain collaboration and logistical service performance. *Journal of Business Logistics, 22*(1), 31.
68. Su, Q., Song, Y., Li, Z., & Dang, J. (2008). The impact of supply chain relationship quality on cooperative strategy. *Journal of Purchasing & Supply Management, 14,* 263–272.
69. Sudolska, A. (2005). Strategiczne partnerstwo jako czynnik umacniania innowacyjności przedsiębiorstw (Strategic partnership as a factor strengthening innovaativeness of enterprises). In Sukces organizacji. Strategie i innowacje, Prace i Materiały Wydziału Zarządzania Uniwersytetu Gdańskiego (Vol. 4, p. 449). Sopot.
70. Surmacz, T., Wierzbinski, B., & Hermaniuk, T. (2013). Collaboration with customers and quality of customer service. In *Servqual as instrument of services improvement and resources management* (pp. 28–37). University of Maribor.
71. Tan, K. H., Wong, W. P., & Chung, L. (2016). Information and knowledge leakage in supply chain. *Information Systems Frontiers, 18,* 621–638.
72. Teece, D. J. (2007). Explicating dynamic capabilities: The nature and microfoundations of (sustainable) enterprise performance. *Strategic Management Journal, 28*(13), 1319–1350.
73. Vachon, S., & Klassen, R. (2006). Extending green practices across the supply chain: The impact of upstream and downstream integration. *International Journal of Operations & Production Management, 26*(7), 795–821.
74. Vachon, S., & Klassen, R. D. (2008). Environmental management and manufacturing performance: The role of collaboration in the supply chain. *International Journal of Production Economics, 111*(2), 299–315.
75. Vázquez, R., Iglesias, V., & Álvarez-González, L. I. (2005). Distribution channel relationships: The conditions and strategic outcomes of cooperation between manufacturer and distributor. *International Review of Retail, Distribution, and Consumer Research, 15*(2), 125–150.
76. Walker, H., Di Sisto, L., & McBain, D. (2008). Drivers and barriers to environmental supply chain management practices: Lessons from the public and private sectors. *Journal of Purchasing and Supply Management., 14*(1), 69–85.
77. Warnier, V., Weppe, X., & Lecocq, X. (2013). Extending resource-based theory: Considering strategic, ordinary and junk resources. *Management Decision, 51*(7), 1359–1379.
78. Wasserman, S., & Galaskiewicz, J. (Eds.). (1994). *Advances in social network analysis: Research in the social and behavioral sciences.* Thousand Oaks, CA: Sage Publications.
79. Wellman, B. (1988). Structural analysis: from method and metaphor to theory and substance. In B. Wellman, & S. D. Berkowitz (Eds.), *Social structure: A network approach.* Cambridge University Press.
80. Wernerfelt, B. (1984). A resource-based view of the firm. *Strategic Management Journal, 5*(2), 171–180.
81. Wierzbiński, B., & Surmacz, T. (2012). Advantages of collaborative approach in customer service management. *Research in Logistics & Production, 2*(1), 115–126.
82. Wierzbiński, B., & Surmacz, T. (2011). Collaboration in supply chains in the process of increasing customer value. In *Management of global and regional supply chain—Research and concepts* (pp. 123–132). Poznań: Publishing House of Poznan University of Technology (Chapter 8).
83. Wincent, J. (2005). How do firms in strategic SME networks build competitiveness? *Journal of Enterprising Culture, 13*(4), 383–408.

84. Wish, M., Deutsch, M., & Kaplan, S. J. (1976). Perceived dimensions of interpersonal relations. *Journal of Personality and Social Psychology, 33*(4), 409–420.
85. Woo, K., & Ennew, C. T. (2004). Business-to-business relationship quality: An IMP interaction-based conceptualization and measurement. *European Journal of Marketing, 38* (9/10), 1252–1271.
86. Yang, Y., & Konrad, A. M. (2011). Understanding diversity management practices: Implications of institutional theory and resource-based theory. *Group and Organization Management, 36*(1), 6–38.
87. Yen, Y. X., & Yen, S. Y. (2012). Top-management's role in adopting green purchasing standards in high-tech industrial firms. *Journal of Business Research, 65*(7), 951–959.
88. Young-Ybarra, C., & Wiersema, M. (1999). Strategic flexibility in information technology alliances: The influence of transaction cost economics and social exchange theory. *Organization Science, 10*(4), 439–459.
89. Zhang, Q., & Cao, M. (2018). Exploring antecedents of supply chain collaboration: Effects of culture and interorganizational system appropriation. *International Journal of Production Economics, 195,* 146–157.
90. Zhu, Q., Sarkis, J., & Geng, Y. (2005). Green supply chain management in China: Pressures, practices and performance. *International Journal of Operations & Production Management, 25,* 449–468.
91. Zhu, Q., & Sarkis, J. (2006). An inter-sectoral comparison of green supply chain management in China: Drivers and practices. *Journal of Cleaner Production, 14*(5), 472–486.
92. Zhu, Q., Sarkis, J., & Lai, K. H. (2007). Green supply chain management: Pressures, practices and performance within the Chinese automobile industry. *Journal of Cleaner Production, 15,* 1041–1052.

Green and Lean Activities of Vertically Integrated Links as a Way of Creating Smart Supply Networks

Anna Maryniak

Abstract In the case of lean supply chains, the main aim is the reductions of costs. With regard to green supply chains, the needs of external stakeholders, including a silent stakeholder—the natural environment, are taken into account. However, currently the co-existence of two types of supply chains is necessary in order for them to meet the expectations of the management staff and the society. Therefore, in this way such supply chains become smart supply chains. In view of the above, it has been assumed that the main aim of the studies is the determination to what extent the studied supply chains are influenced by green and lean ideas within operational and strategic spheres. The studies were conducted on the Polish market, among medium-sized and big production enterprises. On the basis of the studies, it has been determined that supply chains are to a greater extent characterized by lean than green activities within all studied dimensions, i.e. concerning product, logistics, management and relations with suppliers. What is more, the activities within the strategic sphere are more popular than the ones within the operational sphere, with regard to both types of supply chains.

Keywords Green · Lean · Supply chain · Social responsibility
Sustainable development

1   Introduction

Currently, the creation of supply chains which are only cost-saving, flexible, agile, vulnerable, resilient or green is not sufficient to effectively act in the highly-competitive market. Therefore, it is necessary to adopt hybrid strategies, which take into account the co-existence of at least two types of supply chains. That is why it is vital to undertake empirical studies with regard to this matter. On the basis of the world literature review, one could conclude that the researchers are

A. Maryniak (✉)
Department of Logistics and Transport, Poznań University of Economics and Business,
Poznań, Poland
e-mail: anna.maryniak@ue.poznan.pl

© Springer International Publishing AG, part of Springer Nature 2019
A. Kawa and A. Maryniak (eds.), *SMART Supply Network*, EcoProduction,
https://doi.org/10.1007/978-3-319-91668-2_8

147

mostly concentrated on particular types of supply chains. The empirical studies concerning the analyses of hybrid supply chains are rare. Among the works which include the considerations with regard to at least two types of supply chains, the dominating ones are the ones in which the differences between agile and lean supply chains are emphasized. In this work, it has been attempted to present the results of the studies regarding green and lean supply chains. The choice of the hybrid for the studies was influenced by the expectations placed on supply chains by business environment, the expectations of decision-makers managing particular links of supply chains, the previous research experience of the author and the identified research gap in the discussed issue. The creation of seemingly contrary strategies influencing the form of supply chains is a huge challenge for business entities. Therefore, it is valid to answer the following questions:

- do Polish market operators create smart supply networks by implementing lean and green ideas?
- is the range of the implemented „lean" and „green" solutions in supply chains at a similar level within particular dimensions of supply chain management concerning the product and the spheres of its flow?

influenced by green and lean ideas.

2 The Hybrids of Green and Lean Supply Chains as a Way of Creating Smart Supply Networks

Supply chains which assume the co-existence of green and lean ideas as their foundation are one of the ways of creating smart and intelligent networks. This statement can be justified very simply.

Firstly, the results of the studies show that currently the relationships between entities in supply chains are increasing, supply chains are more and more often subject to reconfiguration processes, the number of entities is increasing and new products are more frequently introduced into supply chains [31, 44]. All of this causes the reconfigurations of supply chains which start to create a wider circle of business partners by transforming traditional chains into dynamic networks.

Secondly, smart supply chains are characterized by various attributes. For instance, these supply chains use an intelligent multi-agent system to stimulate supply chain management [25], reduce costs and optimize transport with the use of by Big Data [7], build partnership based on intelligent information requirements [40] and implement a wide range of other activities.

In subject literature, the relationships between the creation of smart supply chains and the idea of lean management [26] as well as the management aimed at the improvement of environmental efficiency [1] are emphasized. The intelligence of smart supply chains manifests itself also in the ability to adjust to the requirements of public opinion and competitive entities in terms of environmental protection and a final level of costs.

3 Characteristics of Lean and Green Supply Chains

The creation of supply chains which take into account the needs of natural environment and which are, at the same time, effective economically is the next step on the way to the evolution of supply chains. According to Nelson et al. [35], after the age of classic supply chains, the interest in lean supply chains had started to increase. Before, enterprises concentrated on the economy of sale, effectiveness and the reduction of operating costs. Waste reduction was attempted not due to environmental motives but the economical ones. However, with time, a greater focus started to be placed on pro-environmental aspects both at the level of manufacturing enterprises and logistic operators [4, 15, 45]. Time and sequential delineations with regard to this matter have always been disputable—that is why the authors usually

Table 1 Supply chain attributes—strategic aspects

Operational aspects	Type of supply chain	
	Lean supply chain	Green supply chain
Variety of products	Low	Moderate, established taking into account easy recovery
Product's life cycle	Long	Closed
Materials	Mainly standard	Mainly innovative
Demand and market segment	Predictable	Selective
Production	Focused on the optimal use of potential	Taking into account the minimization of environmental losses
Stock level	Low	Moderate
Priorities of transport services	Systematic and frequent	Taking into account the minimization of exhaust emission

Table 2 Supply chain attributes—operational aspects

Operational aspects	Type of supply chain	
	Lean supply chain	Green supply chain
Access to the information concerning market needs	Updated with moderate frequency	Updated with moderate frequency
Strategy	Low costs and quality	Quality and innovation
Partnership in supply chains	Traditional partnership	Partnership aiming at transferring green models
Length of cooperation	Mainly permanent	Mainly permanent
Basic indicators	Time, cost, quality	Concerning pollution and usage
Problems connected with supply chain management	Longer reaction time to market changes	Long time of return on green investments and a relatively small segment of recipients

limit their considerations to the characterization of the above aspects, without describing their temporal determinants, since they are aware of the fact that it is hard to reach a consensus with regard to this matter. In Tables 1 and 2, the author has included the basic characterization of the described supply chains, taking into account operational and strategic spheres.

4 Present State of Research

The subject matter concerning lean and green management of supply chains is undertaken in many works included in scientific journals as well as in monographs. The cost-saving topics have been raised by Coimbra [11], Martin [30], Monden and Minagawa [32], Myerson [33], Packowski [36] and Plenert [39], and more detailed works concerning the pro-environmental issues have been published by Emmett and Sood [16], Khan et al. [24], Lyons [28], Sarkis and Dou [43] and Wang and Gupta [46]. However, there are not many papers linking these two subject matters. It is also hard to find books [37] and reports [17]. Kasemsap [23], among others, is of opinion that applying lean SCM strategies and green SCM strategies in the global business environments will significantly improve organizational performance. Kainuma and Tawara [22] stress the fact that a simultaneous improvement of both financial and environmental results is possible. By adopting the idea of the extended supply chain (in which reuse and recycling within the whole product and services life cycle are taken into consideration), they propose the multiple attribute utility theory method for assessing a supply chain. Thanks to this method, it is possible to evaluate the performance of a supply chain not only from a managerial viewpoint but also from an environmental performance viewpoint. Carvalho and Cruz-Machado [8], in reference to the considerations of Kainuma and Tawara [22], have created a conceptual model which, among other things, takes into account the correlations between green and lean supply chains (as well as agile and resilient supply chains). The authors have assumed that the supply chain attributes' values are a consequence of different supply chain practices implementation and they will directly affect the values of supply chain's key performance indicators. Carvalho and Cruz-Machado have adopted "cost", "service level" and "lead time" as key performance indicators to evaluate the effect of each type of supply chain performance.

Ruiz-Benitez et al. [41] have focused on studying the relationship and links between lean, green and resilient supply chain practices and their impact on environmental performance. In their work, lean supply chain practices appear as drivers for green and resilient supply chain practices and their impact on environmental performance is higher than the one of the resilient supply chain practices. Duarte et al. [13] have proposed a conceptual model in which they apply the lean and green measures and adopt the balanced scorecard perspective in order to reach the benefits of supply chain performance. The authors underline that it would be beneficial to know if some lean/green attributes are more important than others with respect to organizational performance.

In contrast, the studies of Zhan et al. [48] show how green and lean practice influences organizational performance and how this association is affected by guanxi (the system of social networks and influential relationships that facilitate business and other dealings). The findings explain that guanxi between organizational partners improves the positive effect of green and lean practice on organizational performance. Duarte and Cruz-Machado [14] have proposed an assessment framework to evaluate businesses in terms of the implementation of a green and lean organization's supply chain. The study reveals that high scores are derived from a good interaction between green and lean implementation in these companies. Garza-Reyes [19] have linked the subject matter of green and lean supply chain management with the issue of total quality environmental management (TQEM). The results proved that, in general, there is less awareness of TQEM in the Chinese manufacturing sector than other environmental and quality/operations improvement approaches such as green supply chain management, reverse logistics, ISO 9000, Six Sigma and lean Six Sigma. Thus, its degree of implementation is also lower than these approaches. Carvalho et al. [9] have concluded that some management practices that are instituted for green or lean benefits have opposite effects on the environmental and economic performance of companies. They have also noticed that not all companies which belong to the same supply chain can be absolutely lean or green. There should be compromises in the individual companies' behavior so that both the environmental and economic constraints of the supply chain are satisfied. Arjestan and Rahimi [3] have studied the relationship of lean production and sustainable supply chains. The authors have demonstrated that if basic elements of lean manufacturing thinking are in line with social factors, it can be assumed that lean manufacturing is a perfect tool to achieve a sustainable supply chain. Bhattacharya [6] has undertaken a similar issue by conducting a systematic review of literature on the integration of lean and green paradigms and their impact on the TBL. As a result of the adopted literature studies, the author has outlined interesting prospective research directions.

The works discussing the above types of strategies concern very narrow issues as well. For instance, Marcilio et al. [29] have analyzed the behavior of environmental and performance variables in the lean manufacturing versus green manufacturing context in a road freight transportation system. The studies concerning transport have also been conducted by Colicchia et al. [12] and Garza-Reyesa et al. [20]. Zagloel et al. [47] have studied the possibilities of implementation of lean rules within reverse logistics. Cherrafi et al. [10] and Nabhani et al. [34] have attempted at studying the links between green activities and one of the chosen lean management tools– lean six sigma. Pandey et al. [38] have developed a framework for the selection of suppliers by evaluating them on the basis of both quantitative and qualitative data, taking into account the pro-environmental and cost-saving aspects. Fercoq et al. [18] have conducted research concerning the lean/green integration, focusing on waste reduction techniques in manufacturing processes.

In general, one may conclude that the topic of lean and green supply chains is discussed in a broader context, i.e. in connection with the issues of:

- sustainable activities, (therefore, not only the pro-environmental ones)
- other types of supply chains,
- other management concepts, e.g. quality management.

    In a narrower scope, the issues regarding both types of supply chains concern:

- fragments of supply chains, e.g. the ones concerning links with suppliers and the bottom-up flow of supply chains,
- specific logistics activities, e.g. transport,
- the chosen tools of lean management.

5   Operationalization and Methodology of Research

Taking into consideration the numerous discussions regarding the definition of supply chain management and logistics Albastroiu and Felea [2], Larsson and Halldorsson [27], Blaik [5], Jeszka [21] and Rutkowski [42] in this work the approach of the so-called unionists, for which logistics is a part of supply chain management, has been adopted.

    Within the framework of logistics, the focus has been placed on operational aspects concerning warehousing, stocks and transport.

    The original definition, in line with which "the simultaneous green and lean supply chain management consists in: designing products and their managing up and down the supply chain by the entities participating in it, with particular focus placed on the need to protect natural environment and minimize all losses" has been adopted in the work.

    Concentrating on both types of supply chains, the analyses within a few basic dimensions have been conducted. The characterization of the studied supply chains has been made in the light of the products moved within supply chains, the implemented logistics activities, the types of links with suppliers as well as the management and configuration of supply chains (Fig. 1).

    The enterprises which took part in the studies were small and big production enterprises, registered in Poland. However, usually their supply chains, on the side of suppliers or recipients, go beyond the country. The sampling frame was

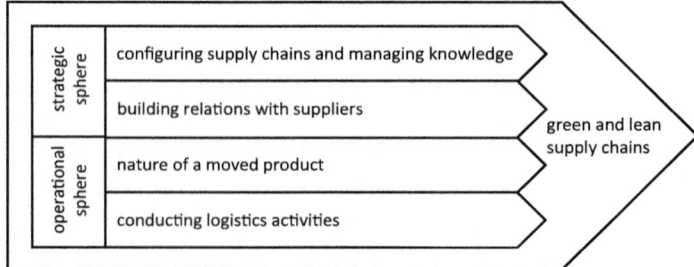

Fig. 1 The dimensions of green and lean supply chains

constituted by the enterprises located in the Wielkopolskie voivodship. The studies were conducted in 2017 by means of an interview, on the basis of a structured survey. In order to obtain high-quality data, the surveys which had not been fully completed were excluded. The total number of enterprises which agreed to take part in the studies amounted to 71 (after the exclusion of 44 defective surveys). The respondents were people responsible for the configuration of supply chains.

6   The Results of the Studies and Discussion

In Figs. 2 and 3, the areas within which the activities concerning lean and pro-environmental strategies are implemented have been identified. Ten test items, which characterize lean and green supply chains, have been attributed to each dimension. The Likert scale has been applied to each question. On the scale, (1) means definitely true, (2) rather true, (3) hard to say, (4) rather untrue, (5) definitely untrue.

Among the activities concerning strategic solutions, the activities which aim at the „at root" elimination of problems occurring during the process of supply chain management are the dominating ones. Taking into consideration the average

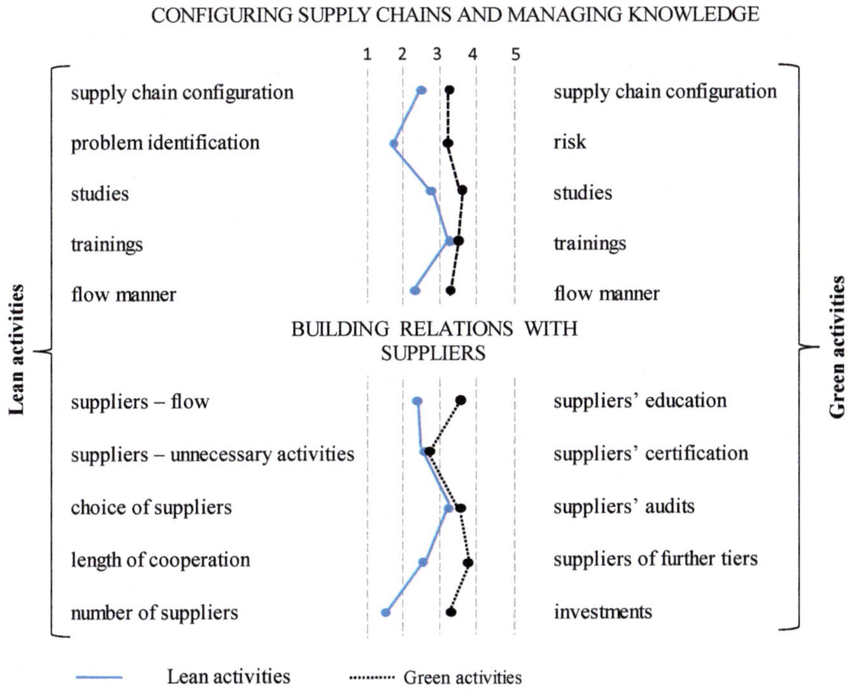

Fig. 2 Lean–green supply chain–strategic dimension

responses from all enterprises, one may conclude that there are no green activities within a discussed area which could be regarded as popular (Fig. 2).

With regard to the links with suppliers, enterprises mainly choose the ones interested in long-term cooperation and they require environmental certificates (Fig. 2).

Within product management, enterprises are focused on lean and pro-environmental activities to a similar extent. For instance, they reduce a number of defective products and strive to eliminate dangerous substances from products/ packagings (Fig. 3).

Among the activities concerning inventory, warehousing and transport, enterprises mainly eliminate unnecessary activities, optimize inventory management, also in the context of its influence on environment, and use environmental-friendly means of transport (Fig. 3).

On the basis of the obtained data, one may conclude that enterprises prove to be more active with regard to lean activities than with regard to pro-environmental activities within all studied dimensions (Table 3).

This may be a result of the mentioned profit-oriented approach, which is the essential goal of the activities of enterprises, as well as the lack of sufficient stimuli from competitive surroundings, which would force enterprises to bear social

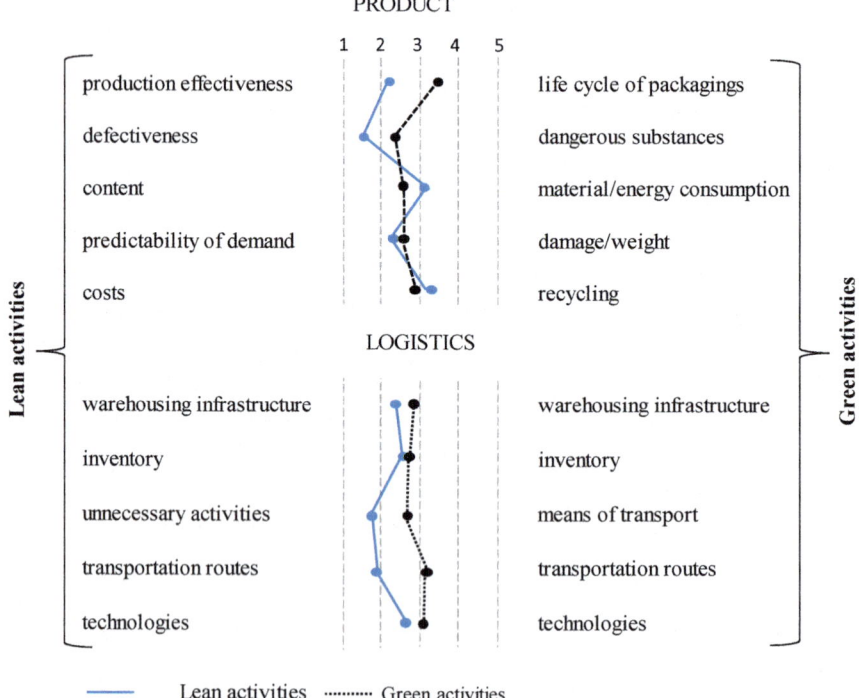

Fig. 3 Lean—green supply chain—operational dimension

Table 3 Lean–green supply chain–strategic and operational dimension

Dimension	Average values for lean activities	Average values for green activities
Configuring supply chains and managing knowledge	2.5	3.3
Building relations with suppliers and recipients	2.6	3.5
Product	2.5	2.7
Logistics	2.2	2.9

responsibility for destroying environmental welfare. As has been demonstrated in the studies [31], the most important factors for business partners in supply chains are speed, punctuality, prices, continuity of supply, payment conditions and trust. The factors connected with environmental protection, for instance a possibility of controlling suppliers of further tiers or a partner's compliance with the environmental law, are of marginal importance.

What is more, enterprises, both with regard to lean and green ideas, are more engaged on the strategic level, than on the operational one. Therefore, it may be assumed that enterprises conduct exploratory, training and research activities which are going to be reflected in business practice concerning logistics (transport, inventory and warehousing) as well as the product in the longer term.

7   Restrictions and Future Research Directions

The presented results of the studies are the exemplification of the outlined subject matter, which requires further research steps. For instance, it is valid to identify consistent, detailed constructs on the basis of the distinguished general thematic groups, which can constitute the basis for the studies of links between the constructs, the determination of power and direction of their relationship and the presentation of the subject matter within the context of detecting causal links within the mutual influences of variables.

According to the studies, Polish market operators create smart supply networks by implementing lean ideas only to a limited extent. This is probably due to the fact that the market in which they operate does not require it yet or these market operators cannot notice clear global trends and they do not see the possibilities of gaining a competitive advantage.

8 Conclusion

On the basis of the studies, one may conclude that within particular dimensions of supply chain management, concerning both the product and the spheres of its flow, the range of implemented „lean" and „green" solutions is varying. The creation of green supply chains is not well-rooted in the Polish culture yet. In contrast, the idea of lean supply chains is definitely more popular, especially within the strategic sphere. In view of the above, it is valid do undertake studies which would reveal in what way the tools of lean management can influence the development of green supply chains.

Thanks to the hybrid created in such a way, supply chains become significantly smarter since they are able to meet the requirements of both the stakeholders directly connected with supply chain links (shareholders, management, regular employees) and the needs of stakeholders which do not have business links with them (society, natural environment). The creation of intelligent supply chains gives greater chances of maintaining competitive advantage.

According to the studies, Polish market operators create smart supply networks by implementing lean ideas only to a limited extent. This is probably due to the fact that the market in which they operate does not require it yet or these market operators cannot notice clear global trends and they do not see the possibilities of gaining a competitive advantage.

References

1. Ahn, K., Lim, S., & Lee, Y. (2016). Modeling of smart supply chain for sustainability. *Lecture Notes in Electrical Engineering, 354,* 269–278.
2. Albastroiu, I., & Felea, M. (2013). Defining the concept of SCM and its relevance to romanian academics and practitioners. *Amfiteatru Economic, 15*(33), 74–88.
3. Arjestan, M. E., & Rahimi, V. (2017). The relationship of lean production and sustainable supply chain by using group TOPSIS method. *Journal of Engineering and Applied Sciences, 12*(8), 2088–2097.
4. Bentyn, Z., & Majchrzak-Lepczyk, J. (2016). In T. Rynarzewski & M. Szymczak (Eds.), *Changes and challenges in the modern world economy: Recent advances in research on international economics & business* (pp. 261–277). Wydawnictwo Uniwersytetu Ekonomicznego w Poznaniu.
5. Blaik, P. (2017). *Logistyka. Koncepcja zintegrowanego zarządzania (The Concept of Integrated Management).* Warszawa: PWE.
6. Bhattacharya, A. (2017). Integrating lean and green and examining its impact on the triple bottom line: A systematic literature review. In *15th ANZAM operations, supply chain and services management symposium* (pp. 1–36). Queenstown, NZ: Lincoln University, June 12–13, 2017.
7. Bucovetchi, O., Simioana, A. E., & Stanciu, R. D. (2017). A new approach in supply chain processes-smart. In I. Bondrea, C. Simion, & M. Inţă (Eds.), *8th international conference on manufacturing science and education—MSE 2017 "Trends in new industrial revolution",* Sibiu, Romania (Vol. 121, p. 07004). June 7–9, 2017.

8. Carvalho, H., & Cruz-Machado, V. (2011). Integrating lean, agile, resilience and green paradigms in supply chain management (LARG_SCM). In P. Li (Ed.), *Supply chain management* (pp. 27–48). New Delhi: InTech.
9. Carvalho, H., Govindan, K., Azevedo, S. G., & Cruz-Machado, V. (2017). Modelling green and lean supply chains: An eco-efficiency perspective. *Resources, Conservation and Recycling, 120*, 75–87.
10. Cherrafi, A., Elfezazi, S., Govindan, K., Garza-Reyes, J. A., Benhida, K., & Mokhlis, A. (2017). A framework for the integration of green and lean six sigma for superior sustainability performance. *International Journal of Production Research, 55*(15), 4481–4515.
11. Coimbra, E. A. (2013). *Kaizen in logistics and supply chains*. USA: McGraw-Hill Education.
12. Colicchia, C., Creazza, A., & Dallari, F. (2017). Lean and green supply chain management through intermodal transport: insights from the fast moving consumer goods industry. *Production Planning and Control, 28*(4), 321–334.
13. Duarte, S., Cabrita, R., & Cruz Machado, V. (2011). Exploring lean and green supply chain performance using balanced scorecard perspective. In *Proceedings of the 2011 international conference on industrial engineering and operations management* (pp. 520–525). Kuala Lumpur, Malaysia, January 22–24, 2011. http://www.commdev.org/files/1339_file_Logistics.pdf.
14. Duarte, S., & Cruz Machado, V. (2017). Green and lean implementation: An assessment in the automotive industry. *International Journal of Lean Six Sigma, 8*(1), 65–88.
15. Dyczkowska, J. (2016). CSR in TSL companies. *Transport Problems, 10*(1), 97–104.
16. Emmett, S., & Sood, V. (2010). *Green supply chains: An action manifesto*. West Sussex: Wiley.
17. EPA. (2000). *The lean and green supply chain: A practical guide for materials managers and supply chain managers to reduce costs and improve environmental performance*. Report: Environmental Accounting Project.
18. Fercoq, A., Lamouri, S., & Carbone, V. (2016). Lean/green integration focused on waste reduction techniques. *Journal of Cleaner Production, 137*(20), 567–578.
19. Garza-Reyes, J. A., Email Author, Y. M., Kumar, V., & Upadhyay, A. (2018). Total quality environmental management: Adoption status in the chinese manufacturing sector. *TQM Journal, 30*(1), 2–19.
20. Garza-Reyesa, J. A., Villarrealb, B., Kumarc, V., & Ruiz, P. M. (2016). Lean and green in the transport and logistics sector—A case study of simultaneous deployment. *Production Planning and Control: The Management of Operations, 27*(15), 1221–1232.
21. Jeszka, A. M. (2014). *Logistyka zwrotów. Potencjał, efektywność, oszczędności (Reverse logistics. Potential, Effectiveness, Savings)*. UE Poznań.
22. Kainuma, Y., & Tawara, N. (2006). A multiple attribute utility theory approach to lean and green supply chain management. *International Journal of Production Economics, 101*(1), 99–108.
23. Kasemsap, K. (2016). The roles of lean and green supply chain management strategies in the global business environments. In J. Sudhanshu & J. Rohit (Eds.), *Designing and implementing global supply chain management* (pp. 152–173). Hershey: IGI Global.
24. Khan, M., Hussain, M., & Ajmal, M. M. (Eds.). (2016). *Green supply chain management for sustainable business practice*. Hershey: IGI Global.
25. Khan, M. Z., Al-Mushayt, O., Alam, J., & Ahmad, J. (2010). Intelligent supply chain management. *Journal of Software Engineering and Applications, 3*(4), 403–408.
26. Kumar, M., Graham, G., Hennelly, P., & Srai, J. (2016). How will smart city production systems transform supply chain design: A product-level investigation. *International Journal of Production Research, 54*(23), 7181–7192.
27. Larsson, P. D., & Halldorsson, A. (2004). Logistics versus supply chain management. An international survey. *International Journal of Logistics. Research and Applications, 7*(1), 17–31.

28. Lyons, K. L. (2015). *A roadmap to green supply chains: Using supply chain archaeology and big data analytics.* South Norwalk: Industrial Press Inc.
29. Marcilio, G. P., Rangel, J. J. D. A., Souza, C. L. M. D., Shimoda, E., Silva, F. F. D., & Peixoto, T. A. (2018). Analysis of greenhouse gas emissions in the road freight transportation using simulation. *Journal of Cleaner Production, 170*(1), 298–309.
30. Martin, J. W. (2007). *Lean six sigma for supply chain management, second edition: The 10-step solution process.* USA: McGraw-Hill Education.
31. Maryniak, A. (2017). *Zarządzanie zielonym łańcuchem dostaw (Green Supply Chain Management).* Poznań: Wydawnictwo Uniwersytetu Ekonomicznego w Poznaniu.
32. Monden, Y., & Minagawa, Y. (2016). *Lean management of global supply chain.* London: World Scientific Publishing Co. Pte. Ltd.
33. Myerson, P. (2012). *Lean supply chain and logistics management.* USA: McGraw-Hill Education.
34. Nabhani, F., Bala, S., Evans, G., & Shokri, A. (2017). Review of implementing lean six sigma to reduce environmental wastes of internal supply chains in food industry. In *15th international conference on manufacturing research, ICMR 2017.* Advances in transdisciplinary engineering (Vol. 6, pp. 327–332). London, UK: University of Greenwich, September 5–7, 2017.
35. Nelson, D. M., Marsillac, E., & Rao, S. S. (2012). Antecedents and evolution of the green supply chain. *Journal of Operations and Supply Chain Management Special Issue,* 29–43.
36. Packowski, J. (2014). *Lean supply chain planning: The new supply chain management paradigm for process industries to master today's VUCA world.* Boca Raton: CRC Press.
37. Palevich, R. (2012). *The lean sustainable supply chain: How to create a green infrastructure with lean technologies.* New Jersey: FT Press.
38. Pandey, P., Bhavin, J., Shah, B. J., & Gajjar, H. (2017). A fuzzy goal programming approach for selecting sustainable suppliers. *Benchmarking: An International Journal, 24*(5), 1138–1165.
39. Plenert, G. (2007). *Reinventing lean: Introducing lean management into the supply chain.* Oxford: Elsevier.
40. Ramanathan, U. (2017). How smart operations help better planning and replenishment? Empirical study—Supply chain collaboration for smart operations. In: H. K. Chan, N. Subramanian, & M. D. A. Abdulrahman (Eds.), *Supply chain management in the big data era. Advances in logistics, operations, and management science (ALOMS)* (pp. 25–49). Hershey, PA: IGI Global.
41. Ruiz-Benitez, R., López, C., & Real, J. C. (2018). Environmental benefits of lean, green and resilient supply chain management: The case of the aerospace sector. *Journal of Cleaner Production, 167,* 850–862.
42. Rutkowski, K. (2004). Zarządzanie łańcuchem dostaw—Próba sprecyzowania terminu i określenia związków z logistyką (Supply chain management—The attempt to specify the term and to determine relations with logistics). *Gospodarka Materiałowa i Logistyka, 12,* 2–8.
43. Sarkis, J., & Dou, Y. (2018). *Green supply chain management: A concise introduction.* New York, London: Routledge, Taylor & Francis Group.
44. Simchi-Levi, D., Kyratzoglou, I., & Vassiliadi, C. (2013). *MIT forum for supply chain innovation, supply chain and risk management.* Massachusetts: Massachusetts Institute of Technology.
45. Surmacz, T., & Fura, B. (2015). Znaczenie ósmej zasady zarządzania jakością w budowaniu zielonych łańcuchów dostaw (The Importance of the Eighth Principle of Quality Management in Building Green Supply Chains). *Logistyka, 2,* 748–754.
46. Wang, H. F., & Gupta, S. M. (2011). *Green supply chain management: Product life cycle approach.* New York: McGraw-Hill Education.

47. Zagloel, T. Y. M., Hakim, I. M., & Krisnawardhani, R. A. (2017). Developing weighted criteria to evaluate lean reverse logistics through analytical network process. In *AIP conference proceedings, 1902(7), November 2017, Article number 020027, 3rd international materials, industrial and manufacturing engineering conference, MIMEC 2017*. Miri, Malaysia, December 6–8, 2017.
48. Zhan, Y., Tan, K. H., Ji, G., Chung, L., & Chiu, A. S. F. (2018). Green and lean sustainable development path in China: Guanxi. *Practices and Performance, Resources, Conservation and Recycling, 128,* 240–249.

The Importance of Information Flow and Knowledge Exchange for the Creation of Green Supply Chains

Anna Brdulak

Abstract Communication in the supply chain is crucial for the smooth functioning of the whole supply chain—raising its value and developing a competitive advantage. The main goal of the article is to show the role of information flows and knowledge exchange for the creation of green supply chains. In her deliberations, the authoress refers to the literature review, to selected case studies and also to her own research carried out on innovative startup projects.

Keywords Green supply chain · Sustainable development · Knowledge exchange Information exchange · Partnership · Cooperation

1 Introduction

The concept of the supply chain came into being in the second half of the 20th century. Multifarious interpretations of the very notion of supply chain should be pointed out. The notion may be considered:

- in subjective terms as "mining, manufacturing, trading and service companies cooperating in various functional areas, and their customers, with product, information and cash flow between all of them" [1],
- in objective terms as two-way streams of material, information and sometimes also finance on the way from raw material mining venues to end customers,
- in the process and activity context as a "combination of processes, functions, activities, relations, itineraries of product, service, information and financial transaction flow inside and between enterprises" [2].

Network perception of the supply chain has recently come to the fore, in connection with its subjective aspect. Hence, the supply chain may be defined, respectively, as "a network of manufacturers and service providers that work

A. Brdulak (✉)
Institute of Logistics, WSB University in Wroclaw, Wroclaw, Poland
e-mail: anna.brdulak@wsb.wroclaw.pl

© Springer International Publishing AG, part of Springer Nature 2019
A. Kawa and A. Maryniak (eds.), *SMART Supply Network*, EcoProduction,
https://doi.org/10.1007/978-3-319-91668-2_9

together to move goods from the raw material stage through to the end user" [3], or as "a network of partners who collectively convert a basic commodity (upstream) into a finished product (downstream) that is valued by end-customers, and who manage returns at each stage" [4]. The network provides for material flows of physical goods, information and capital flows in procurement, production, distribution and customer service.

Currently, against the background of growing environmental issues, it is becoming increasingly important to comprehensively analyze and reduce the environmental impact of business activity taking account of all players of the supply chain. The relationships between the players and their interactions with the natural environment are the focus of attention of green supply chain management. Green supply chain management forms a basis for effective actions to protect the environment as part of mutual cooperation of individual players. At the same time, it is a potential source of development and implementation of eco-innovative solutions in the entire supply chain.

The concept of the green supply chain was proposed for the first time in 1996 by Manufacturing Research Consortium (MRC), Michigan State University, USA. The concept stressed the relationship between the natural environment and optimization of production in the supply chain. Its chief purpose was to minimize the negative environmental impact by tracking and tracing the product throughout its lifecycle, as well as monitoring products after their useful life.

Beamon [5] was one of the first to use the term "green supply chain" in subject literature. The notion was streamlined as a tool for process analysis with reference to the SCOR model by Cash and Wilkerson [6].

The definition proposed by Srivastava was another step in further development of the notion. Srivastava defines the green supply chain as an integration of environmental thinking and supply chain management processes. This refers in particular to processes related to product design, material procurement, production and delivery of the product to the end customer and management of used products whose lifecycle has come to an end [7].

Hence, the fundamental goal of the green supply chain will be to minimize the negative environmental impact of the product at each stage of its lifecycle. This objective will be achieved owing to reverse flows. The focus on reverse flows is the result of the enterprises adapting to change in terms of environmental requirements and carrying on business [8].

The green supply chain is thus a type of response to the growing awareness and ensuing environmental expectations of the stakeholders. By exerting pressure on their business partners, stakeholders attempt to restrict the negative environmental impact of the enterprises [9]. While competing for the customer, enterprises are forced to search for solutions in their business strategies that will not only improve the efficiency of their business, but will also factor in environmental issues.

Businesses strive for maximizing their profits in the entire supply chain, as well as for minimizing their adverse environmental impact. As a consequence, they need to take care of the integration of processes occurring throughout the whole chain.

To this end, they need smooth information flow and the exchange of knowledge and experience between all players of the chain.

2 The Objectives and Management of Green Supply Chains Against the Background of Information Flows

As mentioned before, the main objective of the green supply chain is environmental protection at each stage of the material and information flow cycle. Extensive subject literature and definitions by various authors also indicate detailed objectives apart from the main objective.

Firstly, it has to be pointed out that the concept of green supply chain originates from the concept of sustainable development. From the vantage point of enterprises and their supply chains, the concept of sustainable development means economically rational, socially acceptable and environmentally friendly use of resources in order to maintain environmental development in the long run [10]. However, it is worth noting that the concept of sustainable development is definitely broader and covers environmental, social and economic issues alike. With the clearly delineated goal of the green supply chain being environmental protection and improvement of the environment, the concept is part of green logistics activities.

Tundys [11] refers to detailed objectives. Apart from referring to green supply chain as one where "environmental issues are of primary importance at each stage of the chain", the author also points to the crucial role of management processes, comprising information flow related to the environmental impact and the need to optimize the processes in a resource-saving manner.

Also Baraniecka stresses the importance of cooperation between all supply chain players as regards the processes that take place during the flow of products, information or waste. The author calls the green supply chain the Eco-logistic Supply Chain [12]. The players of this chain include raw materials suppliers, manufacturers, wholesalers, end customers and all entities associated with reverse logistics. They are waste suppliers (processing, manufacturing, extraction and trading companies, as well as consumers themselves), enterprises engaged in gathering, storing, processing or transporting waste, waste recipients and entities engaged in supporting the flow of waste and the accompanying information (logistic operators, organizations dealing with waste recovery), and in regulating and supervising the flows in the green supply chain (state agencies and NGOs). All enterprises are required to liaise closely in order to meet the assumptions of the green supply chain concept. This means i.a. unambiguously interpretable, transparent and smooth information flow, acceptance and standardization of the semantic code deployed for the communication of messages and standardization of specific procedures in case of emergencies.

Srivastava remarks in his definition that the green supply chain concept can be implemented when environmental thinking is integrated into supply chain management, from material sourcing to end-of-life management of the product after its useful life [7] (Fig. 1).

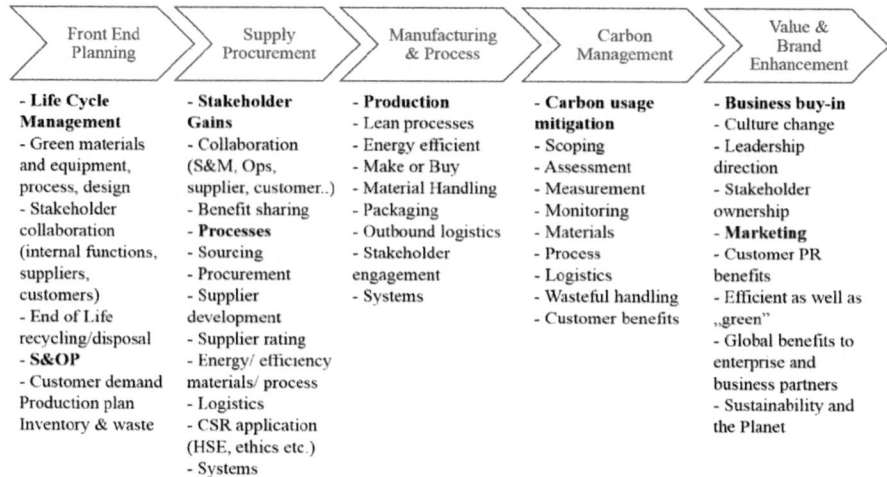

Fig. 1 Own work based on green SCM elements overview [13]

Zhu et al. [14] provide an interesting formulation of the goal of the green supply chain. The authors formulate the goal as a new archetype for companies operating within various supply chains. The achievement of the goals of the archetype is connected with profit increase. This is done by eliminating or reducing the negative environmental impact of processes occurring in individual enterprises to increase the environmental effectiveness of the entities.

It should be stressed at this point that some enterprises perceive consumer pressure on taking environmental action as an opportunity to achieve higher profits. For example, General Motors (GM) managed to reduce its waste disposal costs by 12 million USD as a result of a programme of reusable containers for its suppliers. This would not have been possible without cautious monitoring of the processes in the entire supply chain in response to growing stakeholders' expectations. General Motors might have been less interested in environmental issues, had it achieved record profits. Nevertheless, owing to fierce market competition and the consequential need to reduce costs the company became engaged in environmental issues and used its engagement as an ideal tool to link the improvement of the company's financial standing and meeting current environmental trends [15].

In the context of the quoted definitions which indicate the goal of the green supply chain, as well as raise the issue of green supply chain management (GSCM), it should be explained what GSCM is all about.

Walker and Preuss [16], by referring to Srivastava's definition of the concept, emphasize the aspect of integration of various processes, including production, procurement or distribution processes, which is of key importance for the management process. However, as the authors point out, environmental thinking is what matters most in management. It comprises both internal aspects (organizational factors) and external ones (legal regulations, suppliers, customers, competition).

GSCM consists in providing for multi-faceted dependencies between individual players and the natural environment. This is demonstrated in a comprehensive analysis and limiting the environmental impact of all supply chain players. The subject literature includes multifarious approaches towards the specificity and objective scope of this notion. According to Hervani, Helms and Sarkis, GSCM covers green purchasing, green manufacturing, material management, green distribution and marketing and reverse logistics [17]. In this area, Zhu and Sarkis identify internal environmental management, external GSCM practices (including cooperation with suppliers and customers), raw materials and resources recovery, as well as eco-design [18].

K. C. Shang, C. S. Lu and S. Li. opted for a different division. They identified six GSCM dimensions: green manufacturing and packaging, environmental participation, green marketing, green suppliers, green stock, and green eco-design [19].

Tundys [11] defines supply chain management as a closed cycle related to design, production, packaging, sales, use and finally waste management. The cycle should also take account of storage, transport and information exchange processes. All actions should be integrated, consistent and should provide for environmental protection (Fig. 2).

The table below sums up the different approaches toward green supply chain (Table 1).

Hence, it can be stated that GSCM consists in taking environmental care into consideration in the multi-faceted dependencies between the players. This is demonstrated in a comprehensive analysis and in limiting the environmental impact of all supply chain players. As a result of a detailed analysis and environmental

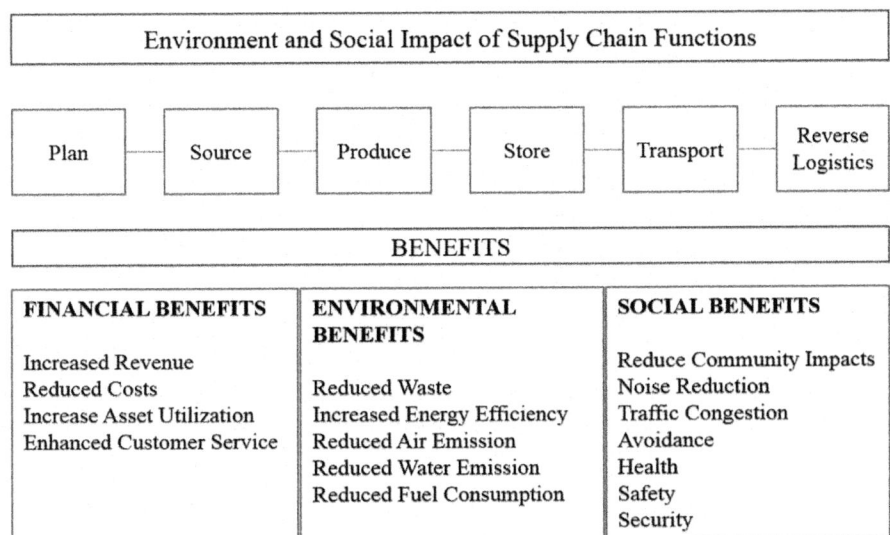

Environment and Social Impact of Supply Chain Functions					
Plan	Source	Produce	Store	Transport	Reverse Logistics

BENEFITS		
FINANCIAL BENEFITS	**ENVIRONMENTAL BENEFITS**	**SOCIAL BENEFITS**
Increased Revenue		Reduce Community Impacts
Reduced Costs	Reduced Waste	Noise Reduction
Increase Asset Utilization	Increased Energy Efficiency	Traffic Congestion
Enhanced Customer Service	Reduced Air Emission	Avoidance
	Reduced Water Emission	Health
	Reduced Fuel Consumption	Safety
		Security

Fig. 2 Own work based on benefits of green supply chain management [20]

Table 1 Green supply chain's approach—literature summary [own work]

Author(s)	Green supply chain's approach
Beamon [5]	The first use of term green supply chain
Hervani et al. [17]	Green supply chain management (GSCM): green purchasing, green manufacturing, material management, green distribution and marketing and reverse logistics
Zhu et al. [14]	The achievement of the goals of the archetype connected with profit increase, done by eliminating or reducing the negative environmental impact of processes occurring in individual enterprises to increase the environmental effectiveness of the entities
Vachon and Klassen [21]	Cooperation of individual supply chain players as part of environmental management
Srivastava [7]	The green supply chain concept can be implemented when environmental thinking is integrated into supply chain management, from material sourcing to end-of-life management of the product after its useful life
Walker and Preuss [16]	1. Integration of various processes, including production, procurement or distribution processes, which is of key importance for the management process 2. Environmental thinking in management—comprising both internal aspects (organizational factors) and external ones (legal regulations, suppliers, customers, competition)
Benaissa and Benabdelhafid [8]	The focus on reverse flows as the result of the enterprises adapting to change in terms of environmental requirements and carrying on business
Shang et al. [19]	Six GSCM dimensions: green manufacturing and packaging, environmental participation, green marketing, green suppliers, green stock, and green eco-design
Flynn et al. [22]	Cooperation of an entity with all supply chain participants as part of intra-organizational environmental management practices (internal integration) and inter-organizational environmental management practices (external integration)
Wu et al. [23]	Cooperation of the functions of an enterprise aimed at engaging employees in the deployment of environmental initiatives and development of environmental skills of an enterprise
Hofer et al. [9]	Stakeholder's pressure to restrict the negative environmental impact of the enterprises
Golinska [10]	The concept of green supply chain originates from the concept of sustainable development
Tundys [11]	The crucial role of management processes, comprising information flow related to the environmental impact and the need to optimize the processes in a resource-saving manner
Baraniecka [12]	Eco-logistic supply chain, cooperation between all supply chain players as regards the processes that take place during the flow of products, information or waste

activities, not only the advantage of one supply chain player, but market force can be referred to; market force results from the deployment of a consistent strategy by all engaged entities.

3 Partnership in Green Supply Chains in View of Knowledge Exchange

Enterprise growth and competitive advantage are possible on condition that an appropriate strategy is implemented, focused on building long-term partnerships based on trust with entities participating in a common supply chain. This is due to the fact that some assumptions related to the sources of competitive advantage for specific sectors (such as price, quality or product portfolio) are similar. Hence, one of the key elements to increase the competitive advantage is to maintain a partnership attitude towards other players in the chain. In this model, apart from gaining measurable benefits such as cost cutting or risk reduction, enterprises liaise even more closely to improve their competitive capability, join complementary competencies or gain knowledge in order to acquire new competencies. Partnership is based i.a. on the ability to liaise with entities operating on the market, to focus on core competencies and to develop all kinds of relations with particular emphasis on customers.

Unfortunately, although partnership in the supply chain in the times of growing uncertainty and dynamic economic change is becoming ever more important, cooperation between enterprises is still difficult to achieve. The main barrier to overcome is lack of trust and insufficient communication, i.e. the exchange of information, knowledge and experiences.

Partnership solutions are still rare in multifarious relationships between enterprises in supply chains not only in Poland, but also globally. Data from 2016 demonstrate that Fortune 500 companies lose about 31.5 billion US dollars annually because they fail to exchange knowledge [24].

According to Bendkowski and Kramarz, transparency of logistics processes in the whole chain is a prerequisite for effective supply chain management. Hence, each player participating in the flow must have access to complete logistic information. Therefore, sales forecast or stock control processes should be considered jointly for all enterprises in the chain rather than from the point of view of a single organization. What appears to be a benefit for one enterprise, does not have to be an advantage for other players within the supply chain [25].

One of the core criteria for an enterprise which attaches considerable weight to the green supply chain is the issue of the extent to which a green approach affects the efficiency of the entire organization and what is the added value that can be gained [26]. In this context, two key groups of management practices can be distinguished: intra-organizational and inter-organizational environmental practices. They are directly associated with green supply chain integration, defined as cooperation of an entity with all supply chain participants as part of intra-organizational environmental management practices (internal integration) and inter-organizational environmental management practices (external integration) [22].

Internal integration revolves around strengthened cooperation of the functions of an enterprise. Such cooperation is aimed at engaging employees in the deployment of environmental initiatives and development of environmental skills of an enterprise [23].

Intra-organizational practices comprise reduced consumption of energy and materials, as well as reduced emissions and waste from internal processes of an organization. In this case, organization and integration of internal management mechanisms is of paramount significance. The mechanisms are expected to bolster communication and increase the employees' engagement, which will ultimately translate into continuous improvement and learning of the entire organization in the environmental field.

It should be stressed that specific actions as part of the strategy to prevent and reduce negative environmental impact pertain to the development of both explicit and tacit knowledge. The learning ability of the organization is also of overarching importance [27].

What definitely matters in this context is a firm's absorptive capacity, i.e. the ability to recognize the value of new information, assimilate it, and apply it to commercial ends [28]. For instance, Lou et al. [29] identify three environmentally friendly supply chains in the Pearl River Delta in China [29]. The amount of environmental losses is estimated based on cost analysis of transport barges and of the use of containers. The information about losses is used to optimize the identified chains.

Internal integration is initiated by the management of the firm and lower-level managers. They support environmental actions by implementing and improving the environmental TQM and environmental management system including environmental review and audit practices, cooperation and knowledge sharing between individual employees and departments within the scope of environmental activities [30].

External integration is based on cooperation of individual supply chain players as part of environmental management [21]. Their cooperation is based on inter-organizational environmental practices. Their application allows for knowledge exchange and creation of social networks, building on trust and engagement of all supply chain players [31]. Partnership pertains to i.a. joint preparation of product lifecycle analyses, eco-design, green distribution and discussing processes within reverse logistics [32]. A good example to illustrate this claim is the way in which the waste management system was implemented in Turkish firms supplying electronics. Andiç et al. [33], who were interested in the potential implementation of a waste management system as a preliminary stage of the green supply chain in the operation of Turkish electronics manufacturers, conducted a survey to this end. The opinions gathered in the surveyed companies confirmed that waste management implementation was potentially useful in the process of green supply chain launch.

Based on the obtained data, a conceptual model was devised to demonstrate the dynamics of the interdependencies between partners in the supply chain with reference to building environmental awareness.

In the quoted example, information flow was efficient and undisturbed. However, this hardly ever happens in real life. In order to reduce the risk related to improper information flow and knowledge exchange in the aforementioned activities, comprehensive product stewardship programmes can be deployed. The programmes focus in particular on cooperation with suppliers and customers in order to solve environmental issues on an inter-organizational and cross-sectional basis [21].

Extended Producer Responsibility (EPR) is an instance of a product stewardship programme. Under EPR, manufacturers and brand owners (called manufacturers) are responsible for the products that they make or sell, as well as for all product-related packaging as soon as it becomes waste. It practically means that manufacturers help cover the costs of collection, transportation, recycling and responsible disposal of these products and materials after their useful life. The EPR concept consists in thinking about individual phases of product lifecycle to hold manufacturers responsible for their products after their purchase and after the expiration of the standard warranty period. Thus EPR helps manufacturers minimize their environmental impact by encouraging them to find ways to reduce costs related to products after their useful life. Better designed products have a longer useful life and afterwards are easier to replace, upgrade or recycle.

Under EPR, manufacturers can also have a greater impact on the operation of product collection and recycling systems. As a result, they have easier access to recycled materials, used in their own supply chains. EPR also has a positive social impact as it promotes fairer and more conscious consumption models. The reason is that EPR costs are frequently included in the product price. Hence, consumers who opt for more expensive products indirectly cover the costs of their disposal later on.

An estimated 400 EPR systems are in operation worldwide (45% of the products and packaging waste are covered by the EPR programme on average in the EU); most of them are mandatory. In the UK, packaging, electric and electronic articles, batteries and cars are subject to EPR requirements pursuant to EU directives (Fig. 3).

It should be stressed that countries like France and Japan have advanced EPR programmes. For instance, there are 14 mandatory EPR systems in France. They comprise additional product streams such as furniture, tyres and infectious medical waste. On the other hand, Japan has extensive legislation which regulates issues relating to product lifecycle in various industries. Some regulations impose the obligation on the manufacturers to apply recycled materials and in new products—parts that are reusable.

A practical example of EPR is the new, all-electric battery-powered Class 8 semi-trailer truck, Tesla Semi, a large and heavy vehicle. What is worth to point out, Tesla Motors designs, develops, manufactures, and sells high-performance fully electric vehicles and energy storage products. It has established its own network of vehicle sales and service centers, as well as Supercharger stations around the world to accelerate the widespread adoption of electric vehicles.

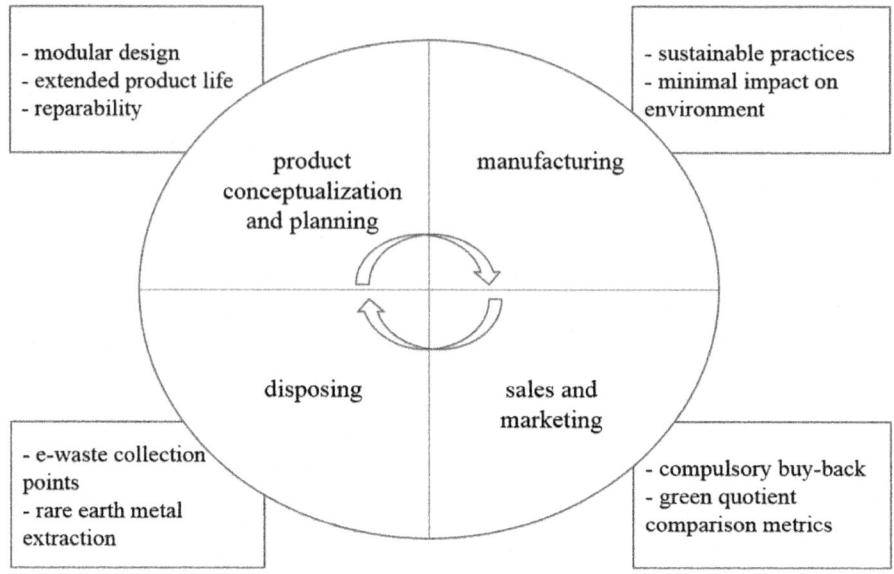

Fig. 3 Own work based on sustainable supply chain model for electronic products [sociallocker]

Its prototype was unveiled in November 2017. Elon Musk used this opportunity to announce a new generation of Superchargers, Tesla Megacharger. The 1000 kWh Tesla Megacharger boasts an impressive range of 400 miles (644 km) per single charge lasting 30 min.

The company claims that the standard version at a price of 150,000 US dollars will have a range of 300 miles, while the upgraded version at a price of 180,000 US dollars will boast a range of 500 miles. Yet, the provided values are subject to change.

Tesla promises superfast solar-powered charging networks all across the US. They will be able to charge the bigger battery up to 80% of its capacity, which will increase the vehicle's range to 400 miles.

Tesla makes an effort to meet the challenges within the green supply chain concept. According to the company's calculations, if a truck driver covers a distance of 90,000 miles annually, and the cost of diesel fuel is 2.70 US dollars per gallon, the cost per 1 mile is 35 cents. The cost of fuel alone will range from 31,000 to 32,000 USD in this case. A battery in a truck will afford transportation companies immense savings.

Moreover, according to an analysis on the Verge website, an average Class 8 truck costs about 120,000 US dollars and uses up diesel fuel worth 70,000 US dollars per year. The cost is ca 0.54 US dollars per mile. Tesla Semi will use less than 2 kWh of energy per mile, so about 26 cents based on the average cost of electric energy in the US. Therefore, Tesla estimates that the offered technology will make it possible to save 200,000 US dollars in fuel throughout the vehicle's useful life. Electric motors are also expected to be much cheaper to maintain than gas engines, and the vehicles' useful life will be longer. In spite of its unquestionably innovative aspect, the proposed solution raises considerable doubts as to its practical application [Green Supply Chain News: Tesla Draws Interest with New Electric Trucks, but Many Raise Doubts it Can Work, [34].

In the near future, the development and implementation of EPR programmes, in particular in the EU states, are expected to be intensified. According to the European Commission, EPR may serve as a useful tool to accelerate the growth of circular economy. As a result, EPR requirements for EU states are likely to become more stringent as part of the Circular Economy Package [35].

4 Practical Application of the Green Supply Chain Based on the Ford Motor Company Solution

The strategy of the global automotive manufacturer, Ford Motor Co., based on the use of recycled materials to make the interior of the car provides an interesting example of an innovative approach towards the creation of a green supply chain. The case was chosen due to the fact, that Ford Motor Co. is one of the market leader in automotive industry, introducing many innovative solutions into manufacturing processes as well as paying special attention to the ecological aspects.

In 2013, Ford announced that the new Ford Fusion would be the first vehicle sold worldwide, with seats made from recycled material and trimmed with fabric using recycled sustainable yarns. Recycled fabrics were already applied in the new Ford Fusion in North America. The company was getting ready to introduce similar solutions in China and in European markets, in Asia and in Europe under the Ford Mondeo brand.

The company was quite adamant about its plans from the very beginning. Ford made the suppliers manufacture ca 1.3 million metres of fabric per year, and each manufactured car contained plastic elements made from a minimum of 40 plastic bottles.

The programme began as early as in 2008, with an idea to make innovative upholstery in the Ford Escape Hybrid. In order to find a textile manufacturer capable of taking on a production order from recycled fabric, Ford had to go beyond the automotive industry. This is currently one of the prerequisites for cooperation with Ford. The company continues to liaise with car seat suppliers using at least 25% of recycled material.

Ford's activity is justified by savings made due to the new solutions. According to the Stanford Recycling Center, recycling of one ton of plastic is equivalent to savings of 5.774 kWh in electric energy. Focus on recycling also stands for reduced waste because waste materials are turned into fabrics. The fabric in the Ford Fusion Hybrid 2013 was made entirely from recycled materials.

According to the representatives of the automotive manufacturer, Ford Fusion is the first global automotive programme in the industry which links recycled fabrics from plastic bottles and industrial waste with car seat manufacturing.

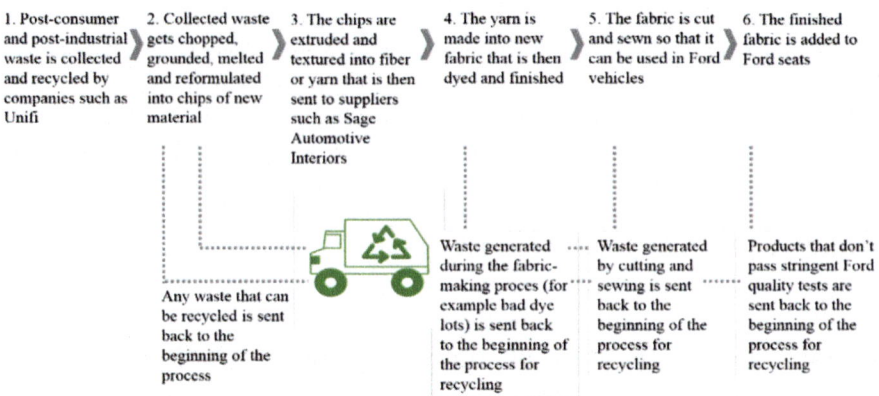

1. Post-consumer and post-industrial waste is collected and recycled by companies such as Unifi

2. Collected waste gets chopped, grounded, melted and reformulated into chips of new material

3. The chips are extruded and textured into fiber or yarn that is then sent to suppliers such as Sage Automotive Interiors

4. The yarn is made into new fabric that is then dyed and finished

5. The fabric is cut and sewn so that it can be used in Ford vehicles

6. The finished fabric is added to Ford seats

Any waste that can be recycled is sent back to the beginning of the process

Waste generated during the fabric-making proces (for example bad dye lots) is sent back to the beginning of the process for recycling

Waste generated by cutting and sewing is sent back to the beginning of process for recycling

Products that don't pass stringent Ford quality tests are sent back to the beginning of process for recycling

Fig. 4 Own work based on closing the loop [36]

Other elements of Ford Fusion also deserve attention. They are seat cushions, seat backs and head restraints containing soy-based foam, underbody components made from recycled battery casings and sound-absorbing material made from recycled cotton. The green supply chain for Ford seats is shown below (Fig. 4).

Unifi is a company which has taken sustainability a step further and developed its own recycled fibre brand called REPREVE, used mainly by textile manufacturers. The fabric is made by chopping, grounding and ultimately melting plastic PET bottles. The liquid obtained in this way is poured through small holes. Next, it cools down and sets in a thread-like shape.

The manufacturing process of the seats begins as soon as Unifi delivers REPREVE yarns to Sage Automotive Interiors, South Carolina. Sage Automotive Interiors uses the yarn to make the right fabric weave, which is then sent to Lear Corporation, Ford's first supplier. The supplier assembles the seats and delivers them to Ford. All production waste, such as bad dye lots or faulty fabrics goes back to Unifi for recycling.

It should be stressed that Ford strives to maintain low costs of raw materials acquisition, while at the same time increasing its employees' commitment to environmental issues. To this end, the company took steps to encourage its staff to collect about 2 million plastic bottles to be used for the manufacturing of REPREVE yarns.

In 2013, Ford liaised with four global suppliers of recycled fabrics. Ford used 41 recycled materials in 15 vehicle models worldwide, from Mustang and Fiesta to F-150 and Taurus, with room for improvement.

A primary element of good and long-term cooperation with supply chain partners is a consistent strategy with calls for environmental care, consistent implementation of the calls, proper information flow and knowledge exchange.

According to Ford representatives, liaising with suppliers was by no means easy at the beginning. Suppliers failed to take the company's guidelines seriously. Under the guidelines, at least 25% of car elements ought to be recycled. However, Ford's consistent search for adequate business partners turned out to be successful. More and more suppliers increased their R&D expenses in the area of environmental protection.

At present, the content of recycled material in each vehicle differs depending on the region, yet it is higher than the initial 25%. In North America it is 100% of the fabric in Fusion Hybrid seats. As regards Ford Mondeo in Asia, the Pacific and in Europe, 43% of the material is recycled.

5 Conclusions

Exchange of information and knowledge in green supply chains is a prerequisite for positive cooperation of supply chain players. The exchange increases the use and improves synchronization of the chain, which is reflected in the benefits derived by each participating entity [37]. Against this background, the green supply chain concept may be perceived as a common network of organizations which strive for environmental protection and which collaborate in order to maximize product value for the end customer. Proper information flow, uniform interpretation of concepts and activities and jointly defined objectives increase the competitive strength of the players who i.a. minimize conflict and focus on the achievement of common strategic goals (see Table 2).

This approach makes the supply chain more flexible and agile by improving the players' ability to make decisions fast [38] and by shortening the duration of individual processes in response to changing customer demands [39]. Sharing knowledge about i.a. product characteristics, processes, the encountered issues and ways to solve them, best practices, concept meanings or information related to the creation of standardized manuals by individual supply chain players increases the value of the entire chain [40].

Table 2 Green supply chain's crucial elements/characteristic [own work]

	Green supply chain's crucial elements/characteristic
Davenport and Prusak [38] Romano [39]	Flexibility and agileability shortening the duration of individual processes in response to changing customer demands
Larsson et al. [40]	Sharing knowledge
Anand and Hanna [37]	Common network of organizations which strive for environmental protection and which collaborate in order to maximize product value for the end customer
Chen (2004)	Various knowledge transfer methods
Sureephong et al. [41]	Positive relations and strategic objectives that are coherent for all entities are of crucial importance
Wu et al. [42]	Close relationship between the partners

It is worth noting that information which is the source of knowledge circulates all over the supply chain and reaches all partners. Hence, each partner controls and interprets the information flow as it sees and understands it. In this way, information obtained from one or several partners is analyzed and transferred to another player in a synthetic form. It must be stressed that each company interprets information in accordance with its own strategy and, depending on the relationship with other entities being part of the chain, passes on communication that has been modified, respectively. Therefore, the quality of the relationship between supply chain players provides a foundation for effective knowledge exchange in the entire chain. In this context, positive relations and strategic objectives that are coherent for all entities are of crucial importance [41].

The paper refers to various practical examples. Clausen et al. [43] also underline the role of information in their research. They focus on the process of configuration and use of transport corridors aided by information. The process makes it possible to achieve sustainability objectives. The authors indicate the advantages of computer-aided systems for itinerary planning, weather forecasts, vehicle tracking and tracing etc., with reference to effective use of forms of transport and supply chain management made easier. The use of these systems makes transport more reliable, less time- and cost-consuming, as well as reduces emissions. The research recognizes the information needs of individual parts of the transport network and proposes the use of relevant information-aided tools for transport processes.

Knowledge transfer may provide long-term competitive advantage in green supply chains [44]. Unfortunately, while it is relatively easy to acquire explicit knowledge, it is much more difficult to transfer tacit knowledge. Various knowledge types require various transfer methods [Chen, 2004]. Again, this is where positive relationships in the supply chain, based on partnership, should be stressed. Owing to a close relationship between the partners, it is easier to share information, also tacit, and interpret it unambiguously, which increases the value of the entire value chain and improves its competitive position [42].

The subject of green supply chains is a development subject. In the context of the conducted analysis, one can point to further directions of research in this area, including analysis of the network of relations between partners, measurement of the increase in the value of the whole supply chain before and after the adoption of a strategy focused on environmental protection or cohesion analysis undertaken in the whole chain of activities.

References

1. Witkowski, J. (2010). *Zarządzanie łańcuchem dostaw. Koncepcje, procedury, doświadczenie* (Supply chain management. Concepts, procedures, experience), Wyd. II zmienione. Warszawa: PWE.
2. Gattorna, J. (2006). *Living supply chain. How to mobilize the enterprise around developing what your customer want.* London: Prentice Hall.

3. Bozarth, C., & Handfield, R. B. (2007). *Wprowadzenie do zarządzania operacjami i łańcuchem dostaw (Introduction to operations and supply chain management)* (p. 30). Gliwice: Helion.
4. Harrison, A., & Hoek, R. (2010). *Zarządzanie logistyką* (p. 34). Warszawa: PWE.
5. Beamon, B. M. (1999). Designing the green supply chain. *Logistics Information Management, 12*(4), 332–342.
6. Cash, R., Wilkerson, T. (2003). *Green SCOR: Developing a green supply chain analytical tool.* LMI-Logistics Management Institute.
7. Srivastava, K. S. (2007). Green supply-chain management: A state-of-the-art literature review. *International Journal of Management Reviews, 9*(1), 53–80.
8. Benaissa, M., & Benabdelhafid, A. (2010). A multi-product and multi-period facility location model for reverse logistics. *Polish Journal of Management Studies, 2,* 7–19.
9. Hofer, C., Cantor, D. E., & Dai, J. (2012). The competitive determinants of a firm's environmental management activities: Evidence from US manufacturing industries. *Journal of Operations Management, 30,* 69–84.
10. Golińska, P. (2014). Metodyka oceny zrównoważonego wykorzystania zasobów w procesach wtórnego wytwarzania – na przykładzie branży samochodowej (Methodology for assessing the sustainable use of resources in secondary manufacturing processes—on the example of the automotive industry). Gospodarka Materiałowa i Logistyka, nr 6/2014 (pp. 17–26). Warszawa: PWE.
11. Tundys, B. (2015). *Zielony łańcuch dostaw w gospodarce o okrężnym obiegu-założenia, relacje, implikacje* (A green supply chain in an economy with circular circulation—assumptions, relationships, implications). Prace Naukowe Uniwersytetu Ekonomicznego we Wrocławiu, nr 383, 785.
12. Baraniecka, A. (2015). *Rozwój ekologistycznych łańcuchów dostaw jako skutek kryzysów: ekonomicznego i środowiskowego* (The development of green supply chains as a result of economic and environmental crises). Prace Naukowe Uniwersytetu Ekonomicznego we Wrocławiu, nr 383, 237.
13. Green supply chain management waste reduction, sustainability and growth. (2015). On-line access: February 12, 2018. aventageconsulting.com/services/green-supply-chain/.
14. Zhu, Q., Sarkis, J., & Geng, Y. (2005). Green supply chain management in China: Pressures, practices and performance. *International Journal of Operations & Production Management, 25*(5), 450.
15. Murray, M. (2017). *Introduction to the green supply chain.* The balance, on-line access: https://www.thebalance.com/introduction-to-the-green-supply-chain-2221084.
16. Walker, H., & Preuss, L. (2008). Fostering sustainability through sourcing from small businesses: Public sector perspectives. *Journal of Cleaner Production, 16*(15), 1600.
17. Hervani, A. A., Helms, M. M., & Sarkis, J. (2005). Performance measurement for green supply chain management. *Benchmarking, An International Journal, 12*(4), 330–353.
18. Zhu, Q., & Sarkis, J. (2004). Relationships between operational practices and performance among early adopters of green supply chain management practices in Chinese manufacturing enterprises. *Journal of Operations Management, 22*(3), 265–289.
19. Shang, K. C., Lu, C. S., & Li, S. (2010). A taxonomy of green supply chain management capability among electronics-related manufacturing firms in Taiwan. *Journal of Environmental Management, 91*(5), 1218–1226.
20. What is green supply chain management? On-line access: February 13, 2018. aims.education/supply-chain-blog/green-supply-chains-management/.
21. Vachon, S., & Klassen, R. D. (2006). Extending green practices across the supply chain: The impact of upstream and downstream integration. *International Journal of Operations & Production Management, 26*(7), 795–821.
22. Flynn, B. B., Huo, B., & Zhao, X. (2010). The impact of supply chain integration on performance: A contingency and configuration approach. *Journal of Operations Management, 28*(1), 54–67.

23. Wu, G. C., Ding, J. H., & Chen, P. S. (2012). The effects of GSCM drivers and institutional pressures on GSCM practices in Taiwan's textile and apparel industry. *International Journal of Production Economics, 135*(2), 618–636.
24. Kowalska, K. (2016). *Wymiana wiedzy w organizacji – jak utrzymać wiedzę i nie tracić milionów?*. February 5, 2018. https://emplo.pl/blog/wymiana-wiedzy-organizacji-utrzymac-wiedze-tracic-milionow.
25. Bendkowski, J., & Kramarz, M. (2006). *Logistyka stosowana metody, techniki, analizy (Logistics applied methods, techniques, analyzes)*. Gliwice: Wydawnictwo Politechniki Śląskiej.
26. Bowen, F. E., Cousins, P. D., Lamming, R. C., & Faruk, A. C. (2001). The role of supply management capabilities in green supply. *Production and Operations Management, 10*(2), 174–189.
27. Gimenez, C., Sierra, V., & Rodon, J. (2012). Sustainable operations: Their impact on the triple bottom line. *International Journal of Production Economics, 140*(1), 149–159.
28. Cohen, W. M., & Levinthal, D. A. (1990). Absorptive capacity: A new perspective on learning and innovation. *Administrative Science Quarterly, 35*(1), 128–152.
29. Lou, Y. H. V., Lai, K.-H., & Cheng, T. C. E. (2013). An evaluation of green shipping networks to minimize external cost in the Pearl River Delta region. *Technological Forecasting and Social Change, 80*(2), 320–328.
30. Apsan, H. N. (2000). Running in nonconcentric circles: Why environmental management isn't being integrated into business management. *Environmental Quality Management, 9*(4), 69–75, 23, 577–591.
31. Shi, V. G., Koh, S. C. L., Baldwin, J., & Cucchiella, F. (2012). Natural resource based green supply chain management. *Supply Chain Management: An International Journal, 17*(1), 54–67.
32. Wong, W. Y. C., Lai, K. H., Shang, K. C., Lu, C. S., & Leung, T. K. P. (2012). Green operations and the moderating role of environmental management capability of suppliers on manufacturing firm perspective. *International Journal of Production Economics, 140*(1), 283–294.
33. Andiç, E., Yurt, Ö., & Baltacıoğlu, T. (2012). Green supply chains: Efforts and potential applications for the Turkish market. *Resources, Conservation and Recycling, 58*, 50–68.
34. Green supply chain news: Tesla draws interest with new electric trucks, but many raise doubts it can work. (2017). On-line access: February 13, 2018. http://thegreensupplychain.com/news/17-11-27-1.php?cid=13362.
35. Perella, M. (2017). *Extended producer responsibility: The answer to cutting waste in the UK?* The Guardian Lab. On-line access: February 12, 2018. www.theguardian.com/suez-circular-economy-zone/2017/may/10/extended-producer-responsibility-the-answer-to-cutting-waste-in-the-uk.
36. Ford motor company keeps pushing the green envelop, as it continues to expand use of recycled materials inside its cars. (2013). On-line access: February 12, 2018. www.thegreensupplychain.com/news/13-05-20-1.php?cid=7093.
37. Anand, B., & Hanna, T. K. (2000). Do firms learn to create value? The case of alliances. *Strategic Management Journal, 21*(3), 295–315.
38. Davenport, T. H., & Prusak, L. (1998). *Working knowledge: How organizations manage what they know*. Boston, MA: Harvard Business School Press.
39. Romano, P. (2003). Co-ordination and integration mechanisms to manage logistics processes across supply networks. *Journal of Purchasing & Supply Management, 9*, 119–134.
40. Larsson, R., Bengtsson, L., Henriksson, K., & Sparks, J. (1998). The interorganizational learning dilemma: Collective knowledge development in strategic alliances. *Organization Science, 9*(3), 285–305.

41. Sureephong, P., Chakpitak, N., Buzon, L., Bouras, A. (2008). *Cluster development and knowledge exchange in supply chain.* arXiv preprint arXiv:0806.0519.
42. Wu, G.-C., Cheng, Y.-H., & Huang, S.-Y. (2010). The study of knowledge transfer and green management performance in green supply chain management. *African Journal of Business Management, 4*(1), 044–048.
43. Clausen, U., Geiger, C., & Behmer, C. (2012). Green corridors by means of ICT applications. *Procedia—Social and Behavioral Sciences, 48,* 1877–1886.
44. Li, S., & Lin, B. (2006). Accessing information sharing and information quality in supply chain management. *Decision Support Systems, 42,* 1641–1656.

Part IV
E-commerce and Digitalization

Mutual Influence of Traditional Trading Chains and E-commerce: Trends and Metrics

Anna Dmytriv and Oksana Kobylyukh

Abstract The aim of this article is to study the tendencies of the transition of traditional trade networks into the e-commerce. On the basis of statistical data, the growing tendency of expanding the trading network of traditional stores through the establishment of online stores has been proved. Research results show that the Ukrainian online trading market is characterized by the development of new product categories such as food and household chemicals. The article provides a schematic list of spheres of e-commerce development influence, which undergo inevitable changes in connection with the popularization of on-line purchases. In this key becomes more actual building of smart supply chains which helps to manage product moving on its way from the production to the finish consumer. When companies are switching from traditional to e-commerce networks, the issue of benchmarking the effectiveness of off-line and online stores appears. In this context, Key Performance Indicators have been analyzed to measure the effectiveness of the online sales and marketing activities of new online sales outlets. The basic system of criteria for evaluating the effectiveness of newly created online stores is proposed.

Keywords E-commerce · Traditional trade chains · Transformation
Key performance indicators · Efficiency

A. Dmytriv (✉) · O. Kobylyukh
Department of Marketing and Logistics, Lviv Polytechnic National University,
Lviv, Ukraine
e-mail: anna.dmytriv@ukr.net

O. Kobylyukh
e-mail: Oksana.Kobylyukh@gmail.com

© Springer International Publishing AG, part of Springer Nature 2019   181
A. Kawa and A. Maryniak (eds.), *SMART Supply Network*, EcoProduction,
https://doi.org/10.1007/978-3-319-91668-2_10

1 Introduction

E-commerce in the mid-1990s was not yet known like a business model, and today it poses a potential threat to traditional trading networks. As one of the authors of the Forbes magazine Steve Olenski wrote, "since Jeff Bezos sent his first online sale through Amazon in 1995, we, consumers, will never go back" [13]. Traditionally, trade has become the most important impetus for the survival of people from the beginning of the history we know, and with the massive spread of the Internet, a paradigm of change in how businesses are doing today.

In his "E-Commerce" book, Ritendra Goel notes that "traditional physical trade in goods and currency is becoming increasingly unpopular and more businesses are jumping on the e-commerce bandwagon" [7]. These trends are also relevant for the Ukrainian market—the border between traditional commerce and e-commerce is becoming more blurry as more and more companies start or continue to integrate Internet technologies into their own business processes. The development of e-commerce issues the growth of necessity to build smart supply chains, which can efficiently manage product movement from the very begging to the finish consumer [10, 12].

There is no a lot of theoretical and methodical surveys about e-commerce in Ukraine. But, the issue of promoting e-commerce in Ukraine, the opportunities and prospects for the transition of traditional trading companies to e-commerce, as well as the emergence of completely new companies that are characterized by only on-line trade, are the subject of research by some Ukrainian scientists. O. Shaleva describes the technology of e-business, its main toolkit and the sequence of actions in the process of organizing e-commerce [22]. As one of the elements of the e-economy, which emerged in the mid-1990s, M. Oklander in his writings recalls e-commerce on digital marketing [20, p. 57]. The questions of the expediency of opening an Internet storefront by a regular store are investigated by Pursky O., Grinyuk B. and Moroz I. [21], also they study the processes of functioning of the trading market on the basis of simulation of competitive interaction between the Internet store and the traditional store [14].

However, in our opinion, it's needed to address a more detailed study to the impact of the development of e-commerce on other sectors of the economy, the potential areas of growth of goods and services for small and medium-sized businesses, the demand for which arises exclusively due to the spread of e-commerce.

2 E-commerce Trends: The Concept of Transformation of Traditional Trade into E-commerce

The widespread use of modern information technology has transformed the Internet computer network into a well-developed infrastructure, and the global Internet network today can be regarded as a huge electronic market that potentially can reach almost all of the world's population. That is why software makers, trade and financial organizations are actively developing various types and ways of conducting business in the global network, which have different character (for example slim or agile) [8, 9]. At the moment, the development of e-commerce has become a very advantageous form of seller relationship with the buyer. The economic properties and features of e-commerce that arose in the process of its formation enabled it to compete in the world of modern global business and created the prerequisites for optimistic forecasts of its future.

Today in Ukraine a large number of traditional stores open their offices on the Internet and become simultaneous participants in traditional and e-commerce. Price differences between this type of sellers and sellers who use only e-commerce are important both from a theoretical and a practical point of view in the light of their impact on the implementation of various elements of the marketing mix. In this context, a large number of questions arise, from which studies were not conducted or were conducted just partly, such as:

- What are the implications for entrepreneurship tendencies for the transition of traditional stores into e-commerce;
- What new, innovative types of goods and services are emerging and developing as a result of increased purchasing of goods through the Internet;
- Which product groups are most popular in e-commerce and where there is potential for growth and development;
- What are the differences in the characteristics and strategies between these two types of Internet vendors, which may have differences in the price differentiation of goods;
- What are the advantages and disadvantages of a competitive trading environment with traditional and e-shops;
- Do sellers can have the advantage of setting higher prices than e-commerce stores using traditional sales channels and e-commerce.

According to Absolunet, a leading American analytical agency in the field of e-commerce in North America, one of the global trends in Internet commerce in 2018 will be the transition of B2B business entirely to online stores [1, 11]. Changing the habits of B2B buyers is that they are already ready to use the features that are typical of B2C-online stores. It is worth noting that large volumes of sales on the industrial market enable B2B companies, having mastered e-commerce, to quickly outperform their competitors. In Ukraine, in 2017, the category of "raw materials and materials" was ranked among the fastest growing categories—according to the number of orders for online growth in the first half of 2017, it was 104% [11] (Fig. 1).

As we see from Fig. 1, the global trend of promotion of e-commerce in the B2B segment is also relevant for the Ukrainian market. It should be noted that this statistics is based on the results of one of the largest e-commerce portal Prom.ua in Ukraine, which represents 93.4 million products from 40.7 thousand verified sellers; The portal turnover is $54.7 million a month, and the monthly customers visit

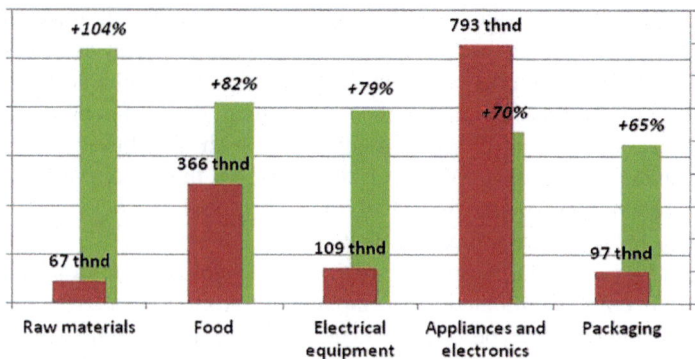

Fig. 1 TOP-5 growing categories in Ukrainian E-commerce market (compare 2017 to 2016 year). *Source* based on [11]

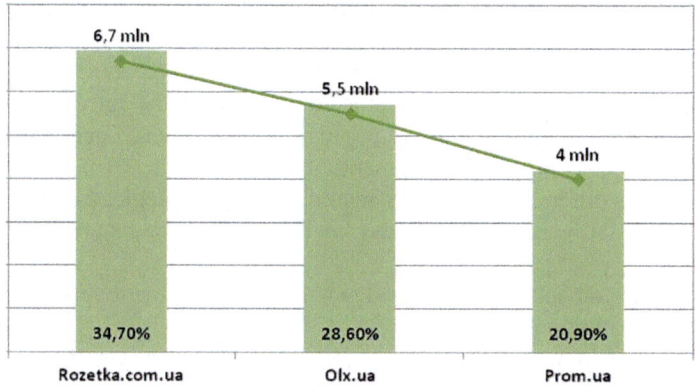

Fig. 2 TOP-3 local Ukrainian E-commerce-resources, most visited by customers in June 2017. *Source* based on [3, 4]

this site 62.8 million times [13]. According to the research company Gemius, this portal is used by 20.9% of Ukrainian users and it ranks third among local Internet resources (Fig. 2). Given the popularity of the portal Prom.ua, we can rely on the information presented in Fig. 1 and assume that the statistics of using the portal and consumer behavior when buying goods of different categories are typical for the whole Ukrainian market.

One of the global trends in e-commerce development in the world is also the duplication of e-commerce—most developed e-businesses with their own online stores also try to have their presence on marketplaces (or, as they are also called, trading platforms) [22, p. 12]. As we see from Figs. 1 and 2—such trading platforms are very popular in Ukraine and they are often used by traditional trading companies for their first entry into the world of e-commerce. Thus, the three largest trading platforms in Ukraine Prom.ua, Olx and Rozetka, according to Gemius

Audience in 2017, reached 53.6% of the Internet population in Ukraine. In general, the Internet audience in Ukraine grew in 2017 from 17.8 up to 19.1 million users. At the same time, the popularity of e-commerce sites increased from 12.3 to 13.6 million people, which is 71% of all Ukrainian Internet users [16]. As for the monetary characteristics of the market, in 2017 Ukrainians spent 14.2 billion UAH on the market, which is 68% more than the previous year, while the number of orders grew by 61% and the average check grew by 4% to UAH 962, which according to the average annual rate of the National Bank of Ukraine equaled 36.17 dollars USA [19].

3 Influence Areas of Transformation Traditional Trade into E-commerce

The development of e-commerce leads to changes in many related areas of the economy. There are a large number of research and articles on the advantages and disadvantages of e-commerce, among which, for example, the ability to make purchases for 24 h, the time lag between the selection and purchase, the availability of complete information on the site of the product, the ability to make a cashless payment, the individual customer care, saving time and money, increased standards of after sales services, leveling of geographic factors, and at the same time the existence of risks of financial fraud, the provision of false information, protection of individuals busty data and so on. The study of general economic phenomena that arise, intensify or level up in connection with the intensification of online trading of companies, today also requires attention from the side of scientists and researchers. The areas of influence of the transformation of traditional trade into electronic can be divided into the following five main groups: consumer behavior, logistics sector, labor market, commercial real estate and IT-sector (Fig. 3). In this study, we will focus mainly on the first two groups, the behavior of on-line buyers and the logistics sector, leaving the other groups as potential for further research.

The behavior of consumers with the onset of the e-commerce era has changed significantly. On-line purchases on the consumer packaged goods market (CPG) in the US over the past 5 years have increased by 350%, and in 2018 amounted to $36 billion, while offline purchases increased by only 3.6% (Fig. 4).

As for Ukraine, for comparison, we can operate data for 2014 and 2017: in 2014, the volume of the e-commerce market in Ukraine was $1.6 billion, and in 2017—$1.9 billion, at the same time, volumes of traditional trade amounted to 36.9 and 22.1 billion USD appropriately, hence the share of online sales in the retail trade turnover of Ukraine increased by more than 4% from 4.3% in 2014 to 8.6% in 2017 (Fig. 5). For data and subsequent calculations, the indicators of the average annual exchange rate of the National Bank of Ukraine were used [19].

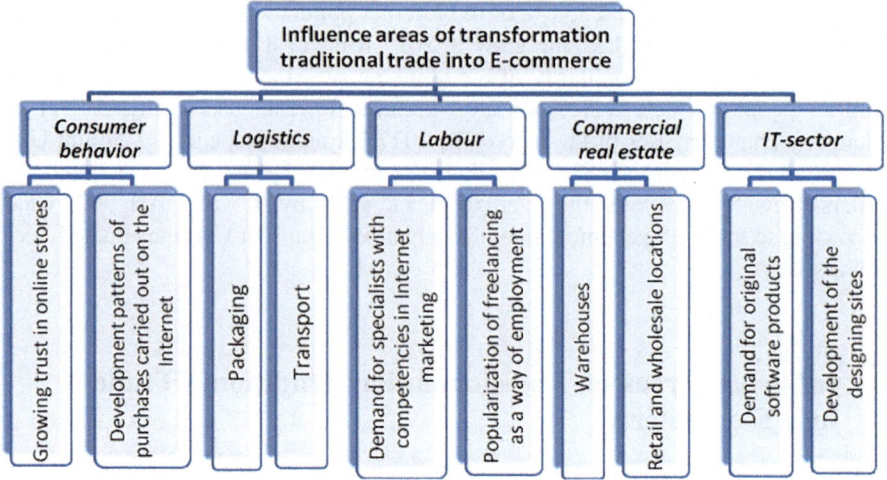

Fig. 3 Influence areas of transformation traditional trade into e-commerce. *Source* own development

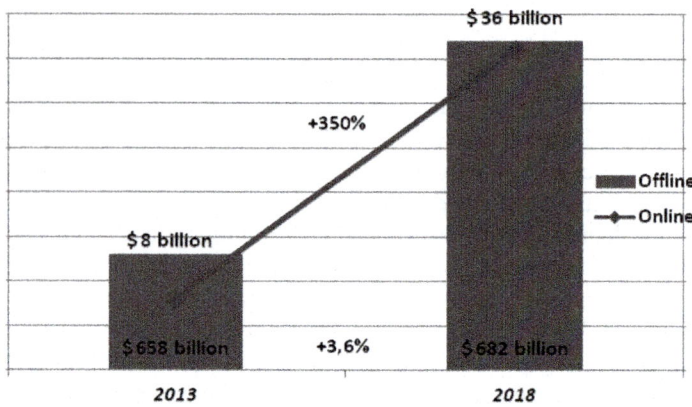

Fig. 4 Online and offline consumer packaged goods (CPG) sales in the United States in 2013 and 2018 (in billion U.S. dollars). *Source* based on [15]

Let's consider the growth rates of offline and online trade in Ukraine. As shown in Fig. 6, the growth of e-commerce in 2017 compared with 2014 was 18.6% in USD, and 165.7% in the UAH equivalents.

At the same time, due to the devaluation of the national currency, with an increase in the total retail trade turnover in UAH equivalents by 34.1%, the dollar equivalent dropped by 40.1% (Fig. 6). Comparing these two indicators, we can conclude that the phenomenon of transformation of traditional trade into e-commerce led to positive growth of e-commerce in dollar terms, against the

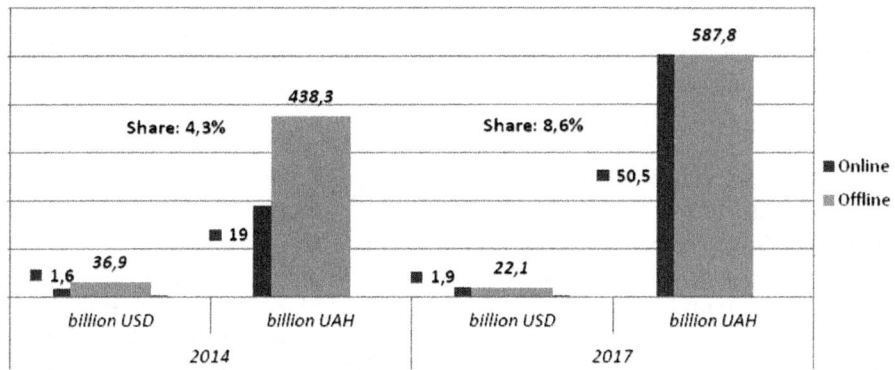

Fig. 5 Comparison of Ukrainian online and offline trade in 2014 and 2017 (share, %). *Source* own development, based on [4, 18]

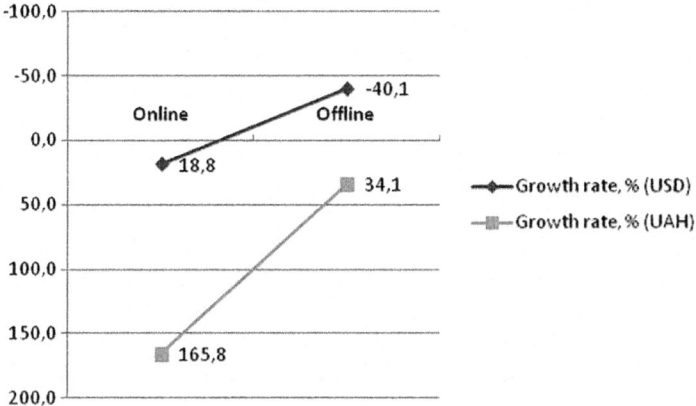

Fig. 6 Growth rates (%) of online and offline sales in Ukraine in 2017 to compare with 2014. *Source* based on [4]

background of decrease of this indicator for traditional trade. We can also make a substantiated conclusion that e-commerce in Ukraine is in a stage of stable development.

Thus, consumer confidence in e-commerce in Ukraine is increasing. Traditionally, Ukrainians buy clothes online, mobile phones, goods for beauty and health, as well as for home and garden. But, there are gradually emerging new online sales areas, including fast-growing food and beverages, tools, hobby goods, and books. This means that demand for goods in these categories will soon increase. Returning to Fig. 1, we can see that in 2017, food orders grew by 82% compared to the first half of 2016, and for the sales leader, the category "Clothing, footwear accessories"—only 32%. The demand for online food purchases has its own peculiarities, Ukrainians often order products of long-term storage or buy certain goods in bulk. During the last half-year of 2017, the highest amounts of food products on the Internet was sold by the following products: tea, coffee and cocoa, chocolate, alcoholic beverages, wafers and cookies, spices, sweets, dietetic products, cheeses, chocolate and nut pasta.

The Internet as a point of sale today is becoming more and more extensive. Yes, companies can either create their own online stores, or use ready-made design developments, or place their offers on a variety of marketplaces under their own brand and adhere to corporate brand strategy. Marketplaces, such as Rozetka, even offer to companies collaboration based on the outsourcing system, to take full care of the organization of the sales process, and to leave for companies only logistical tasks. But in the future such electronic platforms are planning to create their own warehouses, so that in one order it was possible to deliver goods from different manufacturers/sellers. This, accordingly, will entail the development of non-commercial real estate, as indicated by us in Fig. 3. The development of postal services is, on the one hand, the reason for the development of e-commerce, and on the other hand—a consequence. For example, one of the most dynamically developing Ukrainian postal services is the delivery service "New Mail", which can be considered as an engine of Internet commerce in Ukraine.

As mentioned above, the trade areas that are currently active in e-commerce, considering Internet resources as an additional way to distribute goods to the end-user market, include traditional trading networks for food and related products, as well as specialized networks on sale of household chemistry. One of the largest drogeries networks in Ukraine Watsons in May 2016 launched its own online store, built on the Hybris platform. Watsons came to the Ukrainian market in 2006, having acquired the well-known at that time the "DC" network, established in 1993. As for today, Drogerie-network Watsons has more than 440 stores and 25 pharmacies in more than 100 cities of Ukraine, with more than 4000 employees serving almost 3 million monthly customers. For the first month of the work, the online store of this traditional network giant sent only 47 parcels, with the number of on-line sessions reaching almost 150 thousand. This low conversion is mainly due to the fact that at the time of the launch of the store, not all the network products were there not all available products have been made enough description, the information could be incomplete, there was an incomplete functional, which today is important for potential buyers. There was also a need to optimize the Internet store for search engines. By the time of site optimization, the number of products on the site increased significantly—from 3000 SKU at the time of launch up to 12,000 SKU, and also the categorization of goods has been improved. Taking into account the fact that Watsons Company was popular in Ukraine and used the developed CRM system for many years, all clients of the company were informed about new possibilities of making purchases through the Internet store. Along with the implementation of the functionality that helps to increase conversions, expanding the capabilities of the customer's personal office, the Watsons Company has been working on increasing the delivery capabilities, upgrading the Store Locator and product cards, enabling buyers to track the ordering path. Also, a year after the launch of the Internet store, the multimillion function was implemented. All these concerted actions have enabled the company to transfer the power of its network to the e-commerce sector—by the end of 2017, 1.5 million traffic is processed monthly through the online store. The number of SKUs has increased to 20,000, while the online store has products that are not available in offline stores. It is also important to note, when comparing traditional trade and e-commerce for this category of products, that the average check online purchase on the Watsons network is several times larger than the average check in traditional stores. The factors influencing this indicator are, firstly, the option of free courier delivery when purchasing a product for a sum of 400 UAH ($15.04), as well as the tendency to purchase in the Internet-store products in large packages that are inconvenient to carry home by them. So, creating additional convenience for customers in the form of time and efforts saving, the company provides itself with a stable retail trade through e-commerce. According to the results of one and a half years of activity, the online section of drogerie network Watsons has taken a leading position among all the traditional stores of the network.

In the food sector, which is characterized by an increase of 82% in 2017 (Fig. 1), the transformation of traditional trade in e-commerce is also quite active. Networks of food stores duplicate their presence in the field of e-commerce in order to increase the volume of sales of their products by creating additional conveniences for an active group of online buyers, which is constantly growing.

Despite the growth of the e-commerce sector, we note that today's online shoppers are becoming increasingly demanding on the quality of content offered by sellers in on-line stores. In this regard, there is a new global trend called Rich Content, which involves optimizing content and modifying the functionality of the site in order to provide the customer with valuable information and facilitate the purchase process at all stages of interaction with the site. This will allow an online retailer to increase sales, increase conversions and win customer loyalty. All online buyers can be divided into the following three types:

- Browsers (visited the site randomly, out of curiosity);
- Seekers (are ready to buy, are interested in a certain category of goods, but did not select specific attributes or model);
- Searchers (know exactly what they need on the site, they are interested in minimizing the time to search, are characterized by a conversion above the average site).

According to research by PsfWeb, these three types of online buyers are in the following ratio: 75, 15–20, and 5–10% appropriately. The main characteristics of the behavior of on-line buyers are the following:

- desire to place an order anywhere (shop, internet, phone);
- to pay for goods in any way (cash, card, virtual money, bonuses, etc.);
- to get the goods anywhere (store, mail, courier delivery, warehouse, etc.);
- to receive after-sales service or return the goods where it is convenient;
- to get recommendations based on the history of brand interaction.

All these peculiarities of the behavior of online buyers should be considered by traditional business representatives, subject to the decision to enter the e-commerce market. Despite the above statistics, which show an increase in Ukrainians' trust in online purchases, one should not forget that the Ukrainian consumer has a stable habit of buying in offline stores, having the opportunity to check on the spot the integrity and quality of goods.

4   Evaluating the Efficiency of the Transformation of Retail Chains to E-commerce

All owners of online stores are committed to ensuring that their resources work efficiently. But how can you measure the effectiveness of the online store? What indicators of his work need to pay attention to assess how commercially profitable to maintain and develop an online store?

In order to succeed in the field of e-commerce, first of all, it is necessary to be guided by such inviolable trade rules as measuring the performance of the project. Marketing indicators are quite specific measurements, so they are reckoned in the KPI system, which allows you to manage the efforts of marketers [5].

KPI (Key Performance Indicators) is the number or point of data used to measure the effectiveness of achieving any goal. For example, for some online retailers, this may increase traffic by 50% next year. For this purpose, performance may be the number of daily visits or traffic sources (contextual advertising, search engine optimization (SEO) or display advertising, videos from YouTube).

Even at the stage of creating an online store you need to think about how it will be evaluated by its work, so that if necessary, be able to make any adjustments. The list of criteria for evaluation in each case may be individual, but in general it will always include 3 groups of performance indicators—KPI Sales, KPI Marketing, and KPI Service (Customer Service). Indicators of the effectiveness of investment are allocated separately.

The definition of KPI begins with a clear set of goals and an understanding of which business areas affect them. For some purposes, there may be many performance indicators—sometimes even too many—so often their number is narrowed to two or three significant data points. KPI Indicators are the values that most accurately and clearly indicate whether a business moves towards a given goal. Needless to say, performance indicators can and must be different for each goal and task of the online store, whether it is increasing sales, improving marketing or improving customer service.

Here are some examples of goals and related KPIs:

- *Goal* 1—Increase Sales by 10% in the next quarter. KPIs include the amount of daily sales, conversion rates, site traffic.

- *Goal* 2—increase conversions by 2% next year. KPIs include conversion rate, bounce rate at the stage of transition to the trash, associated delivery costs trends, and competitor price trends.
- *Goal* 3—Increase site traffic by 20% next year. KPIs include traffic, traffic sources, ad clicks, social links, bounce rates.
- *Goal* 4—reduce the number of calls to customers by half in 6 months. KPIs include classification of service calls, identification of pages of a site visited directly before a call, events that led to a call.

It is easy to see that there are many performance indicators, and their significance is directly related to measuring the progress of the goal. In the fourth example, monitoring what pages a potential customer visits before calling for support makes sense as it helps identify bottlenecks that, after eliminating, reduce the number of customer calls, but for example 3, this KPI is virtually inappropriate.

As to the benefits of using KPIs and their requirements, it's important to not only formulate and use KPI in their work, but to do it right. Otherwise, the problems will be more than benefits, so the customer's expectations will not be achieved, and the specialist may not receive payment, as the criteria for evaluating his work do not correspond to reality.

Thus, you can formulate the following requirements that KPIs have to meet:

- Transparency—everything must be clear to both the performers and the customer;
- Absence of contradictions—a situation where one indicator is contrary to another, is inadmissible;
- Easy tracking—to fix the score by criteria, you do not need to be an "gurus" of Excel, or spend extra time on the calculations;
- Accounting for the interest of both parties—the positive dynamics of this indicator should be favorable both to the customer and to the agency;
- Measurement of effectiveness—the criteria should give an idea of how the work progresses, which of the channels used is the most advantageous and whether the chosen direction is the right one.

And another important point. There should not be a lot of KPI in SEO, otherwise it just gets confusing. Let there be only 3–5 really important indicators than two dozen not particularly needed.

The idea that KPIs should be based on measurable goals allows you to highlight a set of general performance indicators for the e-commerce industry. Table 1 shows key performance indicators in e-commerce, although, of course, they are not limited to the only ones.

In the following chapters, let's consider the more basic ones.

Table 1 Key performance indicators in e-commerce

Key performance indicators for sales	Key performance indicators for marketing	Key performance indicators for customer service
• Sales volumes (time, day, monthly, quarterly and annual) • Average order size/Average check (sometimes referred to as average market basket) • Average revenue per visit • Conversion rate • The share of abandoned baskets • Comparison of orders from new customers with sales to existing customers • The cost of goods sold • The ratio of the total available market value to the retailer's segment • Affinity of products (which products are ordered together) • Value of products (which products are viewed consistently) • Inventory levels (inventory levels) • Competitor prices	• Output of the web resource in the TOP • Traffic to the site • Conversion • The share of repeat visitors • Refusal to purchase • Thrown baskets • Time spent on the site • Number of page views • Source of traffic • Time analysis (when visitors come during the day) • The number of subscribers to the mailing list • Text message subscribers • Number of initiated chat sessions • Number of subscribers in social networks • Volume of paid traffic • Traffic to a corporate blog • The number and quality of reviews for goods •Branded or display advertising CTR •Indicators of effectiveness of affiliate programs	• Indicator of the cost of customer involvement • Customer return rate • Processing speed of the order • Feedback • Growth of ice base • The number of emails from customers • Number of customer calls • Number of chat conversations with clients • Average time to solve the problem • Classification of issues

Source own development based on [2]

5   Key Performance Indicators for E-commerce Sales

The group of KPI for e-commerce sales includes such basic indicators:

1. *The average purchase receipt* is one of the key indicators in any business. The indicator is calculated as the ratio of the orders number to the amount of revenue of the Internet store for a certain period—month, quarter, year, etc. With these data you can assess the effectiveness of the sales department, the relevance of the system of discounts and in general, the work of the Internet store.

The higher the average purchase check compared to the funds invested in each buyer, the better for the online store, because it means increased profits. If the amount of the average check is almost equal to the cost of attracting the buyer or even below it, then urgent measures to correct the situation—either reduce the cost of attracting buyers, or direct efforts to increase the average check, for example, through cross-selling or up-selling.

It is important not the size but the depth of the average check, because the more items you can sell to the customer, the higher the cost of the check. If it is oriented to the category of light goods, then you can additionally sell it any component from the same category. If the customer buys a fairly expensive luxury product, then you can, accordingly, sell an additional service for its service. That is, the presence of two or more products in check is one of the good indicators of efficiency in e-commerce.

You can increase the average check of an online store by the following methods:

- use the techniques of cross-selling and resale (withdrawal of product recommendations);
- setting the threshold of the cost of the order, after which the delivery will be free of charge;
- introduction of loyalty programs for regular buyers and attractive commercial offers—related products, discounts and incentive bonuses, etc. For example, offer in an online store to buy an additional product at a discount or a general discount for the amount of the order.

2. *Sales*—a criterion for the development of an online store: the more sales are made over a period of time, the greater the sales turnover of the store and the higher the profit in the end. The sales volume is a figure that is assigned to the plan by managers for a certain period of time. Plans can be placed on a month, quarter or year. From the point of view of KPI's online store in the e-commerce segment, the sales plan must be presented in the context of a month, a week, and a day. That is, each manager must daily analyze these targets and compare them with real data. Moreover, in terms of the current traffic that is projected, each manager must know the conversion plan.

Net profit—an indicator for which, in fact, an online store was created. Profit data needs to be analyzed to determine whether the net profit of the owner is satisfied. If the size of this indicator is unsatisfactory, it signals that in the work of the Internet store it is necessary to change something—to reduce costs, to increase the trade margin, to control the relevance of prices, to increase the number of buyers, etc.

3. *Conversion* is one of the most important indicators of this group, if not the most important one. It's important to evaluate not the number of clicks, but the number of targeted actions, that is, sales. The conversion is the number of calls and the number of orders executed, the result of which is the sold product. Conversion—the ratio of the number of buyers to the total number of visitors to the online store—clearly shows how effective the marketing is, whether it leads to the online store of targeted visitors or generates traffic in vain.

If store visitors are targeted, but the conversion is still low, it may be a problem with usability that does not satisfy customers in terms of ordering, shipping and payment, and so on. It should start with the optimization of the most important pages of the site: a commodity card; registration; a basket; ordering you should test the buttons "add to the basket", the registration form on the site, the formats of the description of the goods, options for ordering (from registrations or without it, in 1 click or in several steps). For example, you can experiment with the location of information blocks, changing call-to-action, adding or adjusting existing items. Even the literate processing of text content can significantly affect the conversion rate.

Conversion tracking is one of the most important indicators for online stores. And, in addition to actual sales, you can track the types of interactions like adding and removing a product from the trash, adding products to Favorites, registering users, and more.

According to Nielsen Norman Group, the average return on e-commerce conversion in 2014 was only 3%. Acceptable conversion rate ranges from 1 to 10%. If it is a lower unit—you should look for a reason. In order to increase the percentage of conversion efficiency, you need to analyze the usability of the site, the relevance of prices, the response to the processing of the order.

Improved conversions allow you to make far more profit from a commercial site, with consistent traffic levels. This is especially important, especially if you use pay-per-view advertising channels.

6 Key Performance Indicators for E-commerce Marketing

With the KPI marketing team, you can determine how well you invest in marketing, and make timely adjustments to your advertising campaigns. This group may include the following indicators.

1. *Thrown basket*—the number of so-called dropped baskets clearly demonstrates whether there are problems with the ordering of goods from visitors to the Internet store. If such an indicator is higher than the average for the niche of the online store or there is a tendency to increase it, then it is important to conduct an additional analysis of the Internet store for problems with usability or other barriers to purchase, as well as engage in remarketing.

As practice shows, the rate of refusal to buy at the stage of the formation of the basket, may be up to 70%, depending on the specifics of the business. According to studies by the Baymard Institute, the average is 67.75%. Why is this happening? There can be a lot of reasons:

- the cost of goods in the order does not correspond to the value indicated in the product card. Or, for example, in the basket there is an additional line of delivery cost, which increases the value of the goods and pushes the customer;
- the promotional code for the discount does not work;

- the visitor does not see whether delivery of the goods to his country or region;
- additional costs, such as taxes, appear on the order form;
- insufficient number of payment options for the order;
- technical problems with payment filling.

It is very important to evaluate the rate of abandoned baskets. It is necessary to test new ideas for optimizing the process of placing an order and working with thrown baskets: send letters with a message that the order is not completed and the goods are waiting for the buyer in the basket, to try to get feedback, why the purchase was not made.

2. *The output of a web resource in the TOP* is an important indicator of a properly selected marketing strategy. How difficult it will be to achieve this depends on the qualifications of the selected specialists, the specifics of the business and the employment of niches.

Organic search is one of the main channels for attracting potential buyers to the online store. And in order to promote an online store in one or another search engine, there is usually a set of measures to achieve the highest positions in the search results for certain keywords. Measuring the achieved position with these words is necessary for a proper assessment of the efforts made, making adjustments to the plan of progress.

As many people already know, website promotion requires time. Especially when it is about a young project and a highly competitive niche. Tracking the visibility of the site through the semantic kernel allows you to solve several problems at once.

The first is to evaluate the dynamics of promotion in the early stages of promotion. Traffic from organic search is quite inert and it can take a lot of time before its stable receipt. But Google Search Console's search visibility allows you to track much earlier. If there is a stable growth, then everything is going well.

The second important point is the assessment of the growth prospects. The wider the semantic core of queries that the site displays in the issuance, the more traffic you can get if you tighten them as close as possible to the top 10.

This indicator is also relevant when analyzing individual pages with a good attendance, so that you can understand which occurrences of queries can be further included in the text, and also by which it can be tightened to get even more conversions.

3. *Number of traffic.* To assess the expediency of investing in the promotion of the Internet store and its products in these or other Internet marketing channels, it is necessary to necessarily analyze the amount of traffic generated by each of the channels. You can do this with popular web analytics services such as Yandex. Metrics and Google Analytics, and it's useful to use UTM tags to get the most accurate results for each of your running campaigns.

You will not get anywhere on just one search. Yes, you can achieve excellent sales on sales of "pure organic", through active work on SEO-optimization site.

However, you need to understand that potential customers can be anywhere, and they themselves will not definitely come—you need to go towards them.

Therefore, a specialist who is promoting the site, must track the traffic on each of the channels, measuring and comparing their effectiveness. In itself, this KPI is not, but the increase in traffic for each of the sources, as a result, will definitely have a positive effect on the business of the client. In addition, it will help track any abnormalities quickly and quickly identify which of the channels it relates to, and draw up an action plan to return everything to their places.

Using such a tool for analyzing both Google Analytics and Yandex Metrics, you can find out which promotion channels work best, which resources bring more visitors to the site, how social networks work, and which channels you underestimated before.

What indicators should be monitored:

- How many new visitors have been received for a specific source
- What are the characteristics of behavior they differ (views, percentage of failures, etc.)
- What is the conversion rate for each of the sources
- What is the level of income for each of the channels

It is necessary to track not only the sources of attracting visitors, but their returns. It may well be that the same email-distribution gives you the best conversion and sales figures, and the online store does not pay much attention to this direction.

If one of the channels shows a good engagement of visitors to the site and sales growth—you need to increase the budget for this channel. And if the channel is not effective, it's necessary to minimize costs or optimize the work, for example, removing ineffective keywords from the advertising campaign in Yandex. Direct or to view the seo-kernel on which the site moves.

4. *The average time spent on the site and the average number of pages viewed.* Many online store owners and marketers ignore the estimates of these two criteria, considering they are not indicative. But they allow you to make some useful conclusions about the work of the Internet store. If these indicators are low, then you should evaluate the quality of site traffic. How fast does your site load? What pages are most viewed on the site, and which ones are less? Users are impatient and expect the site to work fast.

It's important to understand that for landing pages or online retailers with a small range these figures can be relatively small, while in large stores, the average number of pages viewed may be more than 20. Also, if the average store of goods in the online store is small in the order, for example, 1–2, then the average time spent by the visitor on the site will be small, because the choice of the product does not need to spend a lot of time.

By analyzing the attendance of grocery pages, you can understand the product benefits of visitors and how they interact with the site. Perhaps the online store has excellent products, but potential buyers cannot find them through poor site navigation. For example, the online store offers good discount products, but they are only available in the general section of the site. It is necessary to create a separate catalog "Sale", "Discounted Goods", "Best Offers", thus attracting attention to the necessary goods.

5. *Refusal to purchase.* In order to understand how effective the pages of the site are we need similar metrics. A refusal is considered to be a variant when the user leaves the site immediately upon leaving the site. With the help of analytics services, you can see the percentage of failure and understand the reasons that caused it. To calculate the failure rate, you need to divide the number of visitors who quickly left the resource by the total number of users who logged in. After analyzing the reasons—design, usability, quality of content, you can reduce the failure rate and increase the flow of targeted traffic.

7 Key Performance Indicators for Customer Service in E-commerce

E-commerce Key Performance Indicators for customer service include:

1. *Indicator of the cost of customer acquisition*—this is the amount that needs to be spent on the client, so that he eventually made a targeted action. This indicator shows the return on investment. Since it is possible to carry out various measures and to invest money in advertising, but if, in addition to clicks, nothing is received, the question of investment payback appears quite acute.
2. *The cost of attracting the target visitor* gives an opportunity to understand how profitable it is to invest in one or another way of attracting customers to the online store. The optimal value of this indicator is strictly individual for each store and even for each specific product group in this store, but in general, the average cost of attracting the buyer should be below the average check, otherwise the continuation of the online store may be simply unprofitable.
3. *The visitor/client's value* is desirable to calculate for each of the traffic channels used, since it allows you to understand whether the investment is paying off and whether it is worth continuing to work in this direction. Moreover, it is necessary to take into account not only advertising costs, but also the cost of the work of a specialist (if he is in the state), as well as the total labor costs.

Often it turns out that the budget for SEO seems to be a great customer, but after calculating this indicator it turns out that this marketing channel is the most profitable. The e-commerce module in Google Analytics is an essential tool for the online store in the context of this goal. The resulting information allows you to reformat the budget, redistributing it to more profitable channels. Also, if necessary, justify the need to allocate additional financing from the customer.

4. *Customer Return/Rebate Indicator.* Many entrepreneurs know that basic sales are not sudden purchases, but customers who are returning for re-purchase. Therefore, the first purchase in this online store should be the best in the life of the buyer—the service; the loyalty system should call back again. If a customer returns for a purchase over and over again, this is an indication that the online store provides quality service.

In addition, it's always easier to retrieve visitors than to attract new ones. To do this, you need to use the tools of collecting email addresses of users, remarketing technology and call-to-action to join company groups on social networks.

Normally, the longer the online store works, the more it has to have returned customers, except when the goods sold by the store are not re-purchased. If the store has been operating on the market for a long time and the audience of its regular customers is still small or absent, this may mean that customers do not find the necessary goods in the Internet store, they do not satisfy the purchased goods or the terms of order and service. Also, a small rate of returning visitors may be in stores that sell only one product for a long time and no options for resale. For example, construction companies: after the purchase of an apartment from a developer, the visitor disappears from the need to visit the developer's site.

5. *Processing speed of the order.* It is important for each order in an online store to track the speed of its processing, since this largely depends on the degree of customer satisfaction with the execution of his order. The optimal order processing time is very individual and depends on both the features of ordered goods and the terms of purchase. Estimate processing speed can be by difference in time of receipt of the order and its coordination with the buyer or, for example, the difference between the time of receipt of the order and its sending.

The faster the application is processed by the client, the higher his response. The reaction to the application is quick—the client is satisfied, he points out the company as a reliable supplier of goods and services. Therefore, it is necessary to use in their work such fast response services as Call-Back.

6. *Feedback.* It is very important to establish feedback with the client in order to understand—how the online store is positioning itself on the market that meets customer expectations or not.
7. *Average time to solve the problem.* If the customer is dissatisfied with the product or services, it is necessary to solve the problem quickly and first of all. So the Internet store will show customers who recognize their mistakes and provide quality service, which in turn protects him from negative reviews on the network, causing confidence in his brand.
8. *Growth of the base of "ice".* If the landing pages (product pages) of the online store are understandable to the users and informative, then it is advisable to methodically inform customers about current promotions and offers by email, leaving the base of the target contacts, this will necessarily trigger the growth of the ice.

Most indicators can be divided into more detailed indicators. Thus, if an online store takes some action to influence Index A, then the situation may arise when A remains unchanged, A1 increases due to impact, and A2 falls due to unclear reasons. The management of the online store concludes that the efforts are not profitable and, as a result of the experiment, the overall indicator A goes down [17]. You always have to clearly imagine what actions are affecting. This is not always the case.

For each business, the indicators will be their own. It is known that new online stores are practically 100% traded without their membership, which means they cannot have criteria related to the warehouse, but on the contrary, they will prevail KPI in assessing the work of suppliers. And in an online store that sells 95% of the warehouse and 5% on-order, the vendor metrics will be completely different. Similarly, different criteria will be used for different product groups. For example, for clothes or shoes, one of the reasons for the failure is "did not fit the size", although for laptops it is excluded.

Taking into account the above, it is expedient to form a basic system of criteria for evaluating the effectiveness of newly created online stores (Table 2).

Table 2 Recommended base system of criteria for evaluating the effectiveness of newly created online stores

Sphere	Indicator	Index content
The site	Traffic (traffic)	Unique visitors during the reporting period
		Total number of site visitors
		Used to calculate the total conversion rate
	Structure of traffic	% specifies how much traffic comes from different channels—search engine, advertising, mailing and direct traffic (traffic sources grow and detail with increasing attendance)
		Is used as the beginning of a detailed analysis of traffic sources
	Number of site orders (cart)	The number of customers who ordered the order through the basket
		Is used when calculating the conversion of the order form, the number of orders through other sources (telephone, ICQ, chat)
	Refusals to the trash (order form)	% of customers who added the item to the basket and then did not make a purchase
		% of customers who started ordering and did not complete
		Both indicators point to disadvantages in the process of purchase: not all information provided, inconvenient form, etc.

(continued)

Table 2 (continued)

Sphere	Indicator	Index content
Sales	The volume of orders	All incoming orders are trash, phone, ICQ, etc.
	Fulfillment of orders	Only shipped orders
	Share of completed orders	% of orders executed in their total number
		Allows you to notice the problem immediately, without using other indicators
	Reasons for unsatisfied orders	The causes may be different: the customer's failure, customer's failure, lack of communication with the client, did not fit the size (clothing/footwear), does not suit the time/rows/methods of delivery
		The reasons for the non-executed orders allow you to immediately determine the direction of further improvements in the work
		The division of responsibility for "causes" between employees gives a positive result
	Number of orders by phone	As a rule, in online stores two sources of ordering—a site and phone
	The share of orders by phone/site	% of orders by phone (site)
		Allows you to determine the total number of orders
	Return of goods	It is used for planning and calculations, in particular for the warehouse. It should be remembered that this is a frozen working capital
	Structure of delivery (in the city by courier/car, in regions, self-service)	Different delivery methods and self-checkout need to be defined to understand the structure and logistics
Call center	Frequency of failures upon receipt (on delivery/self-service)	It is calculated for each type of delivery, taking into account self-delivery
		It is important to take into account the specific reasons for the refusal, for further steps in checking the equipment or changing the delivery partner
	Number of incoming calls	The number of incoming calls to the call center during the reporting period
	Average number of processed calls per manager	The average value of processed calls per manager
	Missed calls share	The growth of incoming and outgoing calls indicates the need to expand the call center or improve the quality of managers

(continued)

Table 2 (continued)

Sphere	Indicator	Index content
Economy	Average service time (sales department)	Average time before the manager is raised (waiting time)
		Average talk time
		Average manager conversion (number of orders to the number of calls)
	Order amount/amount of orders executed (turnover)	The sum of all orders/orders executed (turnover)
		It is appropriate to calculate the reasons for non-fulfillment in the context
	Average check	the amount of orders executed is divided by the number of orders
		It is appropriate to count on the various promotional channels or categories of goods—the average price in the category
	Gross profit	The difference between the net income from sales and the cost of the goods from the supplier
	Average gross profit by order	Gross profit divided by the number of orders executed
		Allows you to understand how best to sell products with a lower margin
	Share of gross profit (% of the amount of orders) of advertising costs	Should be less than 100% at normal work and may be more than 100% when the customer base is consciously formed, with the prospect of higher revenues due to repeated purchases
		It is expedient to count on different channels and different categories of goods
Advertising	Cost of visitor (buyer)	Funds spent on attracting visitors (buyers) are divided by the number of visitors (buyers)
		Possible calculation with the inclusion in the cost of advertisement advertising salaries and SMM-manager
		It is expedient to count on different channels and different categories of goods
		Makes it possible to determine measures to stimulate sales
	The cost of attracting a new buyer/the cost of re-ordering	Usually attracting a new buyer is more expensive than re-buying
		Makes it possible to decide on the increase/decrease of the cost of attracting a new buyer, to expand the client base and future revenue from re-purchases

(continued)

Table 2 (continued)

Sphere	Indicator	Index content
Personnel	The share of orders for each manager	The ratio of the number of orders to a specific manager to the total
		Allows you to evenly distribute the load and signals the need for state expansion or process optimization
		Allows effectively using the system of motivation of the personnel
	The share of wages of staff in gross profit	The salary fund, including taxes, is divided into gross profit
		In effective work there should be less than one (100%)
		The value of the indicator should fall, while hiring new employees to give a slight boost

Source own development based on [2]

At different stages of "maturation" of the online store will require different indicators. In the beginning, general metrics such as a conversion across the site will be analyzed, and you will continue to delve into the traffic source conversion, the cost of attracting a new customer, and the cost of re-purchasing existing customers.

There are still a lot of key performance indicators for the online store. But to track immediately all such indicators are inappropriate and, moreover, ineffective. Therefore, it is recommended to focus on the 5–10 most important indicators for a given task. For example, if there is a problem of increasing the conversion, then the most significant KPI in this case will be the conversion rate and the amount of thrown baskets. But if, say, the goal is to increase traffic, then you need to track the amount of traffic, the performance of different advertising channels, the rate of failure.

After setting goals and defining KPIs, monitoring these indicators should become a daily routine. And most importantly: business decisions should be made on the basis of efficiency, and KPI should be guided by one or another action.

Methods of working with performance indicators are:

1. *Regular monitoring (problem prevention).* When things are going well, orders are pretty good and profits are increasing, as a rule, statistics are thought to last. KPI monitoring helps to spot weaknesses and take action in a timely manner. Configure the analyst in a hurry when troubles have occurred, late. For example, the number of orders increases, the number of returns, too. It's kind of logical. But statistics show that the percentage of returns itself has become higher. It is necessary to understand the reasons. What goods are more often returned? Why? Which courier service has more returns? You may need to change the transport company, refine the product or improve the client service.

2. *Solution to the problem.* Any problem in the online store depends on several factors. Analyze everyone, find a weak spot and fix it. For example, the effect of advertising has fallen. The online store is investing more money, and buyers are no longer getting it. Here you need to analyze KPIs for advertising channels: advertising costs, order count, conversion, average bill. Identify the most damaging channel, or refuse it, or find a solution to how to make it more effective—it's better to focus on advertising messages, change the point of entry or redraw an advertising message. And what about external factors? All KPI depend on many external factors that you cannot influence—seasonality, weather, availability of competitors, time of month, weekends and holidays, unstable political situation, etc. There may be no effect on advertising, as a competitor appeared and took away all the customers. Here it is even more necessary to analyze performance indicators. Identify your strengths and improve them. Influence on external factors is not possible, but to improve the work of the company to you strength. And KPI will help.

3. *Achievement of the goal.* For the purpose, the appropriate KPIs are determined and you need to work on improving each one. For example, the goal is to increase the conversion in the next quarter. You need to analyze the conversion rates for each sales channel and the number of unfinished orders. Incomplete orders are a huge field for improvements. In the analytics system, you can configure the constituent purpose (in Yandex.Metrics) or the sequence of goals (in Google Analytics) for the trash to see at which stage people go. A peculiar crater of sales—it will also allow you to find out which visitors from each of the sales channels are more likely to throw the basket, which goods are most often forgotten. If customers go through the process of entering personal data, you need to improve the relevant page. Maybe it is unclear to them in what form to enter data, or captcha is complicated. If you go from the payment page, it may be worth adding payment methods.

8   Conclusions

The rapid growth of the e-commerce market and the massive transition of traditional commerce to the Internet lead the growth of various tangent spheres of the economy, including logistics. Logistics services with the development of on-line trade have got a significant impetus for development, in particular the Smart Supply Chains have arose, and they ensure the optimization of the processes of moving goods from the starting points of production to the end user through a combination of optimal delivery methods. Further research requires structural changes in the logistics market and Smart Supply Chains that arise under the influence of the transition of traditional business into e-commerce.

The KPI list for e-commerce, which is presented above, is far from exhaustive. However, these indicators, in 90% of cases, will be quite sufficient for the effective development of the online-store; an understanding of what stage is now a business, which marketing channel gives better returns and where to go further.

Performance Indicators (KPIs) can be called a kind of cornerstone on the path to success in online trading. Monitoring Key Performance Indicators will help e-commerce-entrepreneurs to track their progress towards goals in sales, marketing, and customer service.

For a KPI specialist in reports, you can bring the interaction with the client to a new level, and remove a lot of possible questions about the effectiveness of its work. The owner of the online store, in turn, gets a visual picture of how successful the promotion of his project, and how the change in these or other indicators affect the business.

The retail system in the coming years will continue to be formed in the plane of omnichannelity. Omnichannelity—is the reality of retail. Consumers are looking for a product on the shelf to buy it online, and study the product online before going to the store.

While maintaining the importance of offline stores, traditional formats will not be able to avoid digital-transformation. The following recommendations will allow better adapting to changing requests from unique customers:

1. It is necessary to rethink the process of sale and purchase. The online market-place changes consumer habits in the field of shopping. More and more formats are coming to the fore in online and offline trade. In France, the number of points Click and Collect exceeds the number of hypermarkets. In the United States, more and more buyers leave home delivery orders, and 18% of online shopping for everyday consumer goods is a subject to subscription.
2. Recognize the importance of mobile devices in building relationships with buyers. According to the GSMA (Association of Mobile Operators), by 2020, the number of mobile service providers in the world will grow to 5.8 billion, with the largest part of the emerging markets. The ways of communicating with customers should vary according to their habits: modern consumers are insep-arable from their mobile devices—and retailers, and manufacturers do not have to divide them.
3. Offer digital-options in the store. Innovative in-store solutions can both streamline the shopping experience and enrich it with new experiences. Consumers do not want to spend time in queues and prefer to receive infor-mation about products and attractive offers promptly. It is in the interests of retailers to respond in a timely manner to the requests of their buyers.
4. Efficient investment in infrastructure. CRM systems, quick cashier shops, a variety of payment options, and support for continuous communications with mobile devices—this is the future that is available now. By prioritizing, one should keep in mind the peculiarities and expectations of individual purchasing groups.

References

1. Amazon's Third-Party Sellers Had Record-Breaking Sales in 2016. http://fortune.com/2017/01/04/amazon-marketplace-sales/ (07. 02. 2018).
2. Brudan, A. (2013). *The KPI compendium: 20,000 key performance indicators used in practice paperback.* The KPI Institute; smartKPIs.com. March 27, 2013.
3. Daubner, R. (2018). *Consumer trends research with e-commerce data.* https://www.g-casa.com/conferences/shanghai/paper_pdf/Daubner.pdf. (05. 03. 2018).
4. E-commerce in Ukraine. https://ecommercenews.eu/ecommerce-per-country/ecommerce-in-ukraine/#news. (01. 02. 2018).
5. Turban, E., King, D., Lee, J. K., & Viehland, D. (2006). *Electronic commerce: A managerial perspective 2006* (4th ed.). ISBN: 0-13-185461-5. Upper Saddle River: Prentice Hall Publication.
6. Farris, P. W., & oth. (2006). *Marketing metrics: 50+ metrics every executive should master.* In P. W. Farris, N. T. Bendle, E. P. Pfeifer, & D. J. Reibstein (Eds.), Japan: Pearson Education, Inc. Publishing as Wharton School Publishing.
7. Goel, R. (2007). *E-commerce.* New Delhi: New Age International (204pp.).
8. Kawa, A. (2017). Fulfillment service in e-commerce logistics. *LogForum, 13*(4), 429–438.
9. Kawa, A., & Maryniak, A. (2018). Lean and agile supply chains of e-commerce in terms of customer value creation. In A. Sieminski, A. Kozierkiewicz, M. Nunez, & Q. T. Ha (Eds.), *Modern approaches for intelligent information and database systems* (pp. 1–11). Berlin: Springer International Publishing AG.
10. Lefter, V., Roman, C., Sendroiu, C., & Roman, C. (2007). Electronic cost alternatives for e-commerce. *Amfiteatrul Economic, 21,* 79–90.
11. Lindberg, O. (2018). *E-commerce trends 2018: 18 areas that will shape online shopping.* https://www.shopify.com/partners/blog/ecommerce-trends-2018. (05. 02. 2018).
12. Marincas, D. A. (2008). Information system for the supply chain management. *The Amfiteatru Economic Journal, Academy of Economic Studies Bucharest, Romania, 10*(24), 236–253.
13. Official Site of EVO Company. https://evo.company/prom-ua/. (05. 02. 2018).
14. Olenski, S. (2015). *The evolution of ecommerce* (Dec 29, 2015). https://www.forbes.com/sites/steveolenski/2015/12/29/the-evolution-of-ecommerce/#715683687145. (05. 02. 2018).
15. Online and offline consumer packaged goods (CPG) sales in the United States in 2013 and 2018 (in billion U.S. dollars). https://www.statista.com/statistics/419230/us-cpg-online-and-offline-sales. (04. 02. 2018).
16. Prom, Olx и Rozetka охватили более половины интернет-пользователей Украины. https://evo.business/prom-poloviny-polzovatelej/. (06. 02. 2018).
17. Siebrecht, K. (2018). *Every company is an e-commerce company.* http://www.supplychain247.com/article/every_company_is_an_ecommerce_company. (07. 02. 2018).
18. State Statistic Service of Ukraine. http://www.ukrstat.gov.ua/operativ/operativ2017/sr/roz/roz_u/roz1217_u.htm. (05. 02. 2018).
19. XE. The World's Trusted Currency Authority. http://xe.com/currencycharts/. (02. 02. 2018).
20. Окландер, М. (2017). Цифровий маркетинг – модель маркетингу XXI сторіччя: [монографія]/ авт.кол.: М.А. Окландер, Т.О. Окландер, О.І. Яшкіна [та ін.]; за ред. д. е.н., проф. М.А. Окландера. – Одеса: Астропринт (292 с).
21. Пурський, О. І. (2016). Моделювання процесів функціонування торговельного ринку за наявності механізмів електронної та традиційної торгівлі/ О. І. Пурський, Б. В. Гринюк, І. О. Мороз// Економіка розвитку. № 1. С. 83–91.
22. Шалева, О. І. (2011). Електронна комерція. Навч. посіб. – К.: Центр учбової літератури (216 с).

Digital Consumer Needs in Digital Supply Network Creation

Marcin Jurczak and Magdalena Kopeć

Abstract Analysing the processes of globalization, digitalization and virtualization has a substantial impact on shaping present-day supply chains. Alongside these progressing trends, the meaning of supply chains and networks increases, especially in the process of fulfilling the needs of digital consumers. A consumer who uses technology every day to acquise goods and services over the internet. The aim of the article is to present a reflection on the mutual relationship of tools and technologies that shape the consumer's digital needs and capabilities to meet those needs through digital, intelligent supply networks.

Keywords Digital consumer · Consumer needs · Digital supply chain
Sustainable supply chain

1 Introduction

The innovative trends of the digital age conceived a new type of user. The *digital consumer*, being a citizen of the world's internet network, which uses modern-day technology in order to acquire goods and services online,[1] regardless of the time of day or geographical location.

A consumer who actively collects, analyzes and selects information about the offers he is interested in prior to making a purchase decision. While pursuing information online, he often refers to social media platforms in the form of forums,

---

[1]Not all services are available online since they require both parties to present (service provider and service recipient), e.g. haircuts and hair services, dental services, cosmetic services.

M. Jurczak (✉) · M. Kopeć
Department of Logistics and Transport, Poznań University of Economics and Business, Poznań, Poland
e-mail: marcin.jurczak@ue.poznan.pl

M. Kopeć
e-mail: magdalena.kopec@ue.poznan.pl

© Springer International Publishing AG, part of Springer Nature 2019   207
A. Kawa and A. Maryniak (eds.), *SMART Supply Network*, EcoProduction,
https://doi.org/10.1007/978-3-319-91668-2_11

blogs and social channels. Using these sources of information he acknowledges the opinions of others, asks questions and eagerly shares his own experiences with his *online companions*, asks for recommendations and leaves comments [21].

This permanent access to current, specific and most of all—interpersonal sources of valuable information created a new type of market, one being controlled by the digital users demand. A market expected to satisfy a well informed, commercially aware and demanding consumer, who in his decisions considers also social, environmental and economical factors.

Consumer, whose assessment translates into the volume and frequency of purchases, and this in turn affects the functioning of a given supply network and determines the legitimacy of their market existence [20]. Furthermore, considering the conditions of hyper-competition, accompanied by the process of commoditization,[2] modern supply networks must demonstrate great adaptability to maximize consumer satisfaction.

The increased assistance of modern technological solutions enables supply chains to effectively react to market trends, track the progress and further development of consumer needs as well as use market niches and opportunities. A confirmation of these statements can be found in the following chapter, where satisfying consumer needs with the proper use of modern technological solutions constitutes a valuable resource for shaping intelligent supply chains.[3]

2   Methodology

Presented in chapter considerations are based on six-phase cognitive diagram (Fig. 1), which was intended to show reflection at links between tools and technologies used for digital consumer needs and possibilities of these needs through digital, intelligent supply networks.

The first stage study was identification of the needs of digital consumer and motifs for using the technological solutions. In the second stage, the focus of authors was on the processes associated with the creation of chains and supply networks, in particular network and supply oriented technological development in digital reality. In the third stage authors has been studied the interrelationship between the chain and the supply and needs of the consumer.

The next stage of was related to the interaction between the client and the requirements needs for digital supply chains and supply networks. In the fifth stage, these considerations has been extended to include the needs of sustainable

---

[2]Commoditization can be defined as the process by which goods that are distinguishable in terms of attributes (uniqueness or brand) end up becoming simple commodities judged mainly by their economic value (price) [7].

[3]The considerations presented in this chapter refer to the subject of a consumer's digital market as well as to the supply network—which is often interchangeably referred to in literature as supply chains.

Fig. 1 Cognitive diagram.
Source Own work

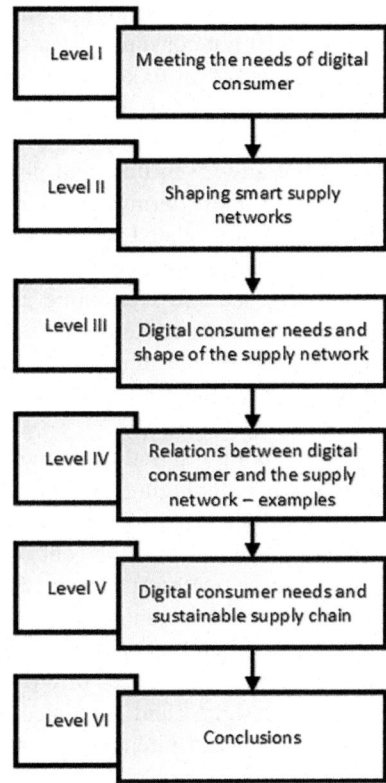

development, both in the context of consumer needs and the supply chains and networks. In conclusion, authors added some trends in modern economy and globalization changes: in both view: for the consumers needs and supply chains and networks.

3 Meeting the Needs of Digital Consumer

Thanks to the technological development and increased information accessibility it can be stated, that the digital consumers needs, a citizen of the World Wide Web, are perceived as global phenomena and trends, that are significantly interrelated.

The needs of a digital, as well as a traditional consumer reflects a perceptible lack of a specific service, product or a physical, emotional, spiritual desire. In order to change this state into a satisfactory experience, a series of incentives and motives enables consumers to act. However, since needs are by definition uncountable, replenishable, complementary, substitutive and tend to reveal themselves with varying intensity—satisfying one need, leads to a emergence of another [22].

According to Accenture's study on "Dynamic Digital Customer" [1] and Ericsson "10 hot consumer trends 2018" [8] the contemporary digital consumer is highly interested in tools and technology based on and related to artificial intelligence, augmented and virtual reality. The customer of today, searches for solutions allowing him to eliminate the boredom of common, everyday life—available on demand at any given time and place.

However these terms are increasing in popularity, the definition of artificial intelligence and related products remains very wide. According to literature, the term artificial intelligence essentially covers everything from automating processes such as e.g. learning, reasoning and auto-correction to actual robotics [17]. At its core, the A. I. can be described as vast sets of data, being appropriately processed and analysed by machines and computerized systems in order to simulate human intelligence by adapting relevant, individually tailored offers to the changing needs of digital consumers.

Contemporary digital consumers regularly experience the use of artificial intelligence while indulging in common day activities, i.e.: "50% of all consumers interact with their service providers through live chats or mobile messaging apps on a monthly basis, and 85 percent of those say it feels like it's easier to get in touch through these methods. These live chats and messaging apps are increasingly supported by AI-driven chatbots" [1, p. 4].

Augmented and virtual reality are examples of technologies further enhancing the day to day experiences of a digital consumer. By definition, virtual reality is a computer designed and simulated, three-dimensional environment, enabling users for movement and interaction, resulting in the stimulation of one of the 5 human senses [3, 11]. Topic related studies show, that digital consumers adopt different types of virtual reality, in areas such as games and entertainment, education, tourism, etc. [3].

On the other hand, augmented reality is a type of technological experience, connecting both worlds—real and artificial. Augmented reality enhances the real world with computer generated content, which is further complemented by computer graphics [3, 12]. The applications of augmented reality in a digital consumers activities can be found in fields such as transportation, sports, product simulations, entertainment, games and other [3].

In comparison to traditional buyers, the digital consumer demands an entirely different market which needs to be constantly updated and modified in order to be more involving and diversified in order to heighten the user experience (Fig. 2). In order to meet these demands, supply chains show high organizational efficiency and an overwhelming rate of introducing technological novelties to the market. It can be stated that the needs of digital consumers are a constant variable, with a disposition for rapid change.

Among numerous demands of a modern-day digital consumer, one needs to include optimizing information availability and transparency, as well as increased data security. While living in the age of information, users express the need to independently manage their personal data, by knowing exactly what information is being shared, shared with whom and for what purpose [1]. This need clearly comes

Fig. 2 The main reasons for buying a smartphone device. *Source* Own work based on Ref. [1, p. 10]

Table 1 Digital consumers motive of using technology based goods and services

Technology	Motive of using technology
Artificial intelligence	– Availability corresponding to the expectation of convenience expressed consumers – Saving time thanks to efficient service and short response time – Personalization corresponding to the expectation of individual treatment
Augmented and virtual reality	– Search for new experiences and stimulus – Trying, simulating the use of products and services – Personalization—both realities allow users to generate an image of a tailor-made product

Source Own work based on Refs. [1, 3]

from a lack of trust in entities operating alongside the supply chain and related networks, as well as the impossibility of controlling them.

The intention of one's right to regulate the flow of their personal information can be summarized in the statement that the "means of stimulating and directing behavior of a human being to satisfy a given need stemming from an unmet need" [22].

Serving as a supplement to the previous considerations is the presentation of research results conducted for the purposes of this publication, which aspires to identify the current motives for using presented technological solutions by digital consumers (Table 1).

The presented intentions of a digital consumers use of technological solutions are a manifestation of certain attitudes that, although repetitive and predictable, are difficult to aggregate into one key area—which is personalization. This becomes evident when compared to the fact that each of these technologies satisfies different consumer needs.

The motive of personalization has, occurred in both the artificial intelligence and augmented and virtual reality technologies. In artificial intelligence, the personalization theme occurred mainly in the context of automating the process of targeting individual consumer characteristics and sharing an appropriate message (offer) to the buyer. However, within augmented and virtual reality the personalization theme occurred mainly in the context of creating a simulation of using a specified product.

It is worth emphasizing that the broadly understood personalization, matters mostly when it accurately recognizes and meets the needs of the digital consumer and this in turn depends on the transparency of information provided by him [1].

4 Shaping Smart Supply Networks

Over the years, the common understanding of supply chains in terms of a supplier, producer, seller and consumer has grown vague. At present, many supply chains are networked, since supply chains connect, and overlap with each other, and individual links become interrelated suppliers, receivers, cooperators as well as competitors [15]. In part, this is due to the following reasons.

First of all, the current driving force behind almost every delivery network is no longer the producer, as it was in the 1970s and 1980s, nor the salesman with whom suppliers fought for the best possible product display during the 1990s and the beginning of the 21st century. Contemporary supply chains are driven by consumer demand, which inevitably led to creating the consumers market [17].

Secondly, the increase in digital connectivity and technological capabilities further influenced the digital consumer's evolution. The modern consumer uses technology on daily basis in order to purchase goods and acquire services provided over the internet, without regard for the time of day or night distinct from location—factors which greatly limit traditional vendors. This is undoubtedly a huge challenge for the functioning of supply networks, since delivering the product to the final recipient is its prime objective.

Furthermore, the digital consumer's market in particular, requires broadly understood transparency, both on the side of demand and supply. From the market demand side, consumers expect transparency in order to learn everything there is to know about the individual attributes of the offer. On the other hand, producers demand transparency from their clients, since the supply side depends greatly on customizing it's offer according to the information provided by individual consumers. Information transparency not only favors the process of deepening relationships, but also allows for adequate adjustments by each member affiliated with digital consumer market. Mutual cooperation on the market, enables modern supply chains to evolve and better accommodate to the rapidly changing user demands.

Lastly, each entity included into the logistics of customer service has smaller or greater, direct or indirect impact on the consumer's final product experience. Therefore, an intelligent supply chain should be composed only of links that affect consumers by creating additional product value [5]. Each link contributes to the total value of user experience through supportive products and/or complementary services, while the consumer provides value in the form of revenue and information streams. This issue is of great importance, since the functioning and organization of entire supply chains is highly dependent on meeting the expectations of demanding consumers.

Progressing developments the field of information technologies, such as cloud computing, mobile technologies and the Internet of Things are undoubtedly conducive to the process of transforming traditional supply networks into intelligent supply networks. A brief description of selected technologies is presented below.

Cloud computing is the solution that contributes most to the process of transforming traditional supply networks into intelligent supply networks. This is due to the fact, that nearly all subsequent technological solutions, i.e. mobile technologies, are being transferred to and further developed in the *cloud*. Cloud computing is a tool that can be identified with outsourcing in the field of Information and Communication Technologies (ICT), allowing external entities for unrestrained, remote use of IT services. The scope of outsourcing although very wide, is determined by three basic service models available in the cloud: Software as a Service, Platform as a Services and Infrastructure as a Service [20].

The Software as a Service (SaaS) model is based on remote access to software over the internet, whose components are located not with the client but on the server infrastructure of the service provider [20]. The Platform as a Service (PaaS) model depends on providing and entire online working environment accessible online. In this case, all software licenses needed by users, can found on the service provider's server [4]. The Infrastructure as a Service (IaaS) model consists in providing the entire IT infrastructure in the form of: servers, operating memory, disk space, software licenses, etc. [20].

Mobile technologies, that do not exist without cloud computing, are becoming increasingly important in the functioning of supply networks. Their use ensures almost immediate availability of information in the system, as they connect to the network and are equipped with appropriate wireless communication modules (radio and optical) [20]. These technologies, further paired with automatic identification and management systems, became an essential tool for improving logistic processes in supply networks [6].

5 Digital Consumer Needs and Shape of the Supply Network

There is a clear link between the needs of the digital consumer and organisation of the supply network. To be able to analyze those links, it is necessary to ask the fundamental question at the beginning: does the shape of the supply network derives from consumer needs? Or maybe the opposite is true: that the creation and support of consumer need is the result of a logistics network? Considerations on this subject is contained in the further part of this section. Whatever the interpretation, the relationship between the needs of the digital consumer and organization networks should be considered in terms of the relationship. The relationship that can be dealt with on many levels, consist of multiple threads and many aspects that affect the environment. And the environment should also be analyzed in several ways: as a business environment, social environment or sustainable development.

Looking at these relationships from the first perspective, it would be necessary to consider in what way digital consumer needs affect the shape of the logistics network. And the impact of this may be different, related to the needs of the consumer concerning the way of providing the information, how meet the consumers need for knowledge, the flexibility of the supply network and the impact, which consumer gives for supply network shape. In turn, taking the opposite assumption, that this supply network has a direct impact the consumer needs, you can put the three main questions:

- how the digital consumer's needs are created through the smart supply networks?
- how the digital consumer's needs are shaped through the smart supply networks?
- how the digital consumer's needs are fulfilled through the smart supply networks?

An example of such activities should be creating special supply chains and networks to support e-commerce. This causes the "creation" for purchasing online on the side of the consumer. By creating new tools, e-commerce, there is a alternative to traditional sales network and the consumer begins to feel the need to purchase the products in a different way than before, although de facto using the same network or point of sale.

In the market there is a lot of companies, who are under the influence of the modern consumer needs, and because of his needs have changes their supply networks. This includes actually all of the distribution networks, not only those companies, which sells today using e-commerce channel. The modern consumer expects, inter alia, the full information. This forced the merchants to reorganization of all the information management process. This means in the first place: information about the product and its parameters, and the second: the availability of this product. This in turn forced on companies creating tools for sharing information about inventory via the Web. The consumer is expecting today the full information about the availability of products in the chosen time horizon, which give him the ability to precisely determine the time of delivery (usually from the time of completion of the contract and the time of delivery, and the unavailability of a product in a natural way this time increases). Some vendors goes even further by providing on its website specific information about the availability of selected assortments in definite retail stores. That, depending on the model of sales organization, allows you to: booking the product in the selected location or it can be for the consumer just a visual information, and as a source of information about the stock in definite, specific stores. Such information could become a base for the next feature for consumers: notification of the availability of products in the central store or select retail locations.

Need for actual information accompanies the digital consumer in the beginning phase: the whole sales process and also later, during the entire delivery process. This clearly affects the entire supply network, by a very important need for providing information about the next stages of delivery: begin of completion process, final of completion process or preparation of order for dispatch. The next step is often to take a shipment by a third party operator, and the need for information takes over the specialized TSL trader (courier, postal operator or other company from a wide logistics services industry). And it's through the tracking system (track and trace) of this operator, digital consumer receives further information.

At the stage of order processing digital consumer can actively participate in creating and shaping the whole delivery process. Depending on the offer a specific company it can mean: the choice of provider of the shipment, selecting the type of service or choosing a specific service (for example: delivery at a specified time). There are also a whole range of services that provides the possibility of shortening the way of delivery—for example dedicated services through special maintenance-free, self-service points (such as packages machines of Inpost) or a wide range of services with the reception points (wide range of services operated by Polish Post, including points in newsstands, petrol stations, etc.). A special offer have also a specific sellers and resellers, even if a special points for personal collection of deliveries (this points can be located also in "traditional" trading points).

How the digital consumer needs impacts on the network and supply chains and organization of logistics processes? Information need forces companies to computerization of processes. This helps to manage and distribute data. Access need: consumer need for having an access to a different number of sales channels forces greater flexibility. In turn, along with the need for quick preparation of the order appears the need to raise the effectiveness of the intralogistics business processes on the provider (seller, logistics provider). This in turn forces the logistics continuous improvement: in the area of services quality, indirectly affecting the shape of specific processes of the logistics service provider (transport, storage, value added services, etc.). In direct terms, it's not always about the process itself, but, above all, information, transparency of actions or update the status of an order or shipment. On this all elements overlaps also the presence of the wide range of mobile tools—e-commerce distribution channel increasingly accompanied by the channel associated with the sale through mobile devices. It is called also m-commerce market, and this m-commerce as a natural subspace e-commerce market is characterized by intense increases. M-commerce adoption is also high among e-commerce merchants, as 80% of e-commerce retailers with at least $50 million in annual sales either currently offer or plan to offer m-commerce. This makes sense given that these companies have already heavily invested in digital, so mobile would be the next logical step for them [16].

6 Relations Between Digital Consumer and the Supply Network—Examples

As an example of how the digital consumer's needs are related to the method of carrying out purchase transactions are intertwined with the way the supply networks are organized, it can be used Allegro, one of the leaders of the e-commerce market in Poland, for many years know as a leader in an auction site for individual users and sales platform for stores and professional sellers; and Inpost—one of the independent postal operators on the Polish market. From research indicated by Allegro, for as much as 69% of respondents, delivery costs have a significant impact on the frequency of online shopping, and up to 83% of online buyers would increase the scale of purchases, with a free return option included in price [13]. The Allegro and Inpost partnership was planned as strategic and long-term. It assumes introduction of services dedicated to Allegro in the Inpost portfolio, among others Allegro Polecony InPost service (Allegro Registered Letter Inpost) or Allegro Paczkomaty InPost service (Allegro Package Machine Inpost). And one of the new opportunities that consumers have been prepared is, among others free return option—which is a valuable alternative to other delivery methods (courier parcels or Polish Post postal and courier services offer).

Digital consumer is expecting today flexibility. The traditional courier service carries restrictions, even in terms of delivery in a specific dates, time or location. Meeting the digital consumer needs comes out today also with self-service points offer. Among them are for example self-service "paczkomaty" points—stationary "package machine" points for courier parcels pickup. The study commissioned by the Inpost company shows that 74% of consumers so far, strongly prefers "package machines" than traditional courier service [14]. This clearly shows what the digital consumer is expecting today. The necessity to wait for the courier shipment, in the era of wide access to the Internet network and popular mobile solutions, is as little discomfort. In particular, that because of access to information, the consumer on a regular basis keeps track of the status of his shipment and delivery. He is informed about its occurrence at the expected time of delivery. And actually, the "machine point" is usually accessible and available 24 h a day, which also affects the increasing flexibility what comes against the expectation of the digital consumer.

Manufacturing companies and commercial face an important question concerning how the organization of the distribution network should be prepared, especially when distribution network is dedicated for a e-commerce consumer. Support for e-commerce market can be regarded in three ways: as part of the "classical" distribution network, as a part of modified distribution network and processing orders with special e-commerce restrictions. An alternative is to create a completely new, parallel distribution network, based on the assumption that the needs of electronic trade (and therefore the digital consumer using for example a mobile) are different.

In Supply Chain 4.0, supply-chain management applies Industry 4.0 innovations—the Internet of Things, advanced robotics, analytics, and big data—to

Table 2 Relations between Industry 4.0, Supply Chain 4.0 and Consumer 4.0

Industry 4.0 and Supply Chain 4.0	Supply Chain 4.0 and Consumer 4.0
Automated production at factory	Predictive shopping
Autonomous truck to warehouse	Shipment rerouting by customer
Automated warehouse	Last-mile delivery

Source Own work based on Ref. [2]

Table 3 Selected needs of digital consumer and challenges for supply network

Need of digital consumer	Challenge for supply network
Information about the products and their availability	Ensuring the tools to track inventory and to share an information about inventory using dedicated tools and channels of communication
Information about the order and delivery	Ensuring the need to track the completion of orders and delivery on the various stages of delivery
Flexibility of delivery	Access to flexible logistics services, along with the ability to change for example the date or time of delivery
Self-service delivery	Access to flexible logistics services, along with the ability pick up the delivery using dedicated infrastructure

Source Own work

jump-start performance, and customer satisfaction [2]. Supply Chain 4.0, as a derived of Industry 4.0, must therefore face challenges such as production automation, automation of warehouse processes, autonomous vehicles of different types including cars, trucks etc. Separate challenges concern the consumer directly: especially in the area of the orders deliveries. This includes the possibility of ordering via mobile phones, order management in accordance with the requirements of the digital consumer (both as a direct supply and deliveries to stores) as well as the challenges of supply within the framework of the last mile (Table 2).

The digital supply chain, as we envision it, consists of eight key elements: integrated planning and execution, logistics visibility, Procurement 4.0, smart warehousing, efficient spare parts management, autonomous and B2C logistics, prescriptive supply chain analytics, and digital supply chain enablers. Companies that can put together these pieces into a coherent and fully transparent whole will gain huge advantages in customer service, flexibility, efficiency, and cost reduction; those that delay will be left further and further behind [19] (Table 3).

A separate group of challenges are those associated with ideas for use in supply network augmented reality or artificial intelligence. Multithreaded processing of large data sets or algorithms support are only few certain areas of supply chain digitalisation. A separate issue is the use of information systems not only to the current support processes and management processes, but also in the area of planning, forecasting or optimization, and therefore wherever it is necessary to use artificial intelligence. Area of knowledge associated with artificial intelligence is an area of very rapid development, a development accompanied by continuous increase in the possibility of artificial intelligence tools. Increasingly, artificial intelligence is used to solve complex decision-making problems in the supply chain.

Similarly with the virtual reality—a few years ago about her presence in the supply chain were only examples. Today more and more bravely talking about using them in your daily work. Google's Alphabet relaunches smart glasses for business use. Google Alphabet is running tests of an updated Google Glass at various corporations, including Boeing, General Electric and Volkswagen. After failing to successfully market the Google Glass to consumers, Alphabet is now aiming at the business world in hopes that the Glass will be better suited to functional as opposed to recreational use. The original Glass cost $1500, but according to the Journal's report, the new Glass price will vary based on company customization and needs [18].

7 Digital Consumer Needs and Sustainable Supply Chain

In the days of ever-shrinking resources and increasing environmental awareness, you cannot omit the trend of creating more sustainable supply chain. How to achieve sustainability in the supply chain? Experts Ernst & Young represent in this area the six basic applications:

(1) supply chain sustainability can no longer be ignored,
(2) companies are predominantly risk driven with aspirations to unlock strategic opportunities and benefits,
(3) companies tailor their approaches and governance to create sustainable supply chains,
(4) leading companies are establishing a shared commitment with suppliers,
(5) technology enables visibility and influence beyond Tier 1,
(6) collaboration is critical for companies to achieve greater business and societal impacts [9].

How to define a sustainable supply chain today? As a derivative of the elements such as:

I. Strategic considerations:

  a. Organisational strategy
  b. Supply chain strategy and structure
  c. Marketing strategy

II. Decisions at functional interfaces

  d. Product design and product life-cycle
  e. Pricing and valuation of returns
  f. Forecasting, information provision, and value of information

III. Regulation and government policies:

    g. Extended producer responsibility

    h. Cap and trade programs

IV. Integrative models and decision support tools [10].

Sustainable supply chain is therefore one that uses appropriate technological tools to ensure the effectiveness of the action: maximize efficiency, achieve the required level of flexibility and minimize the impact on the environment. It is at the same time, the supply chain based on strategic decisions about organizational structure or marketing strategy, based on the right product life cycle, pricing and financial strategy. And one of the important things for sustainable supply chain is also an external impact and support—appropriate policy and regulations.

And only remains the question: how to reconcile the challenges arising from the implementation of the sustainable supply chains with the needs of a changing world and the requirements of the modern, digital consumer? The answer to this question is contained in the consumer's digital needs and how to meet these needs with today's delivery network. Last mile delivery increasingly are using fleets that use alternative drives. This allows you to minimize the impact on the environment for the supply, which is seen above all in urban areas. The multitude of ways of delivery provides the flexibility, so much desired by the modern consumer. This flexibility, understood as a possibility as the most precise fit how to deliver supplies to the consumer needs, allows you to maximize the efficiency of the supply. In turn, maximizing the efficiency of the supply directly translates to maximizing the use of resources (fuel, different types of assets), which again has a positive impact on the supply chain and its overall, global impact on the environment.

The consumer is becoming today more and more aware and take decisions bearing in mind not only their economic situation, but also the impact of the decision on the environment. Access to modern technologies, changes in approach to purchases or new communications capabilities—all of this beneficial effect on the consumption of resources of the environment. Delivery through specialized entities of the industry often means giving up the traditional purchasing process, which reduces unnecessary movements of the consumer, reduces energy consumption, and finally—in line with the philosophy of sustainable development. With the delivery of a specialized entity may receive the additional expenditure "environmental"— even if packaging necessary in the process of delivery. Usually, however, such materials and raw materials that can be recycled.

8 Conclusions

Globalization and digitalization are changing surrounding reality. They are changing consumers, networks and supply chains. Digital consumer analyzes data, compares products and deals, optimizes his purchasing process. This is reflected also in the network and supply chains, which also are forced to compare, analyze, and optimize. Digital consumer creates a new market in which traditional and modern sales channels operate parallelly and stores are accompanied with their virtual counterparts. Modern technologies are starting to dominate in all parts of the supply chains. They are becoming an integral part of the sales process and delivery process. New technical solutions and organizational support transformation in network and supply chains—from traditional to digital.

Along with the evolution of the supply chains and networks consumer needs are also changing. The need for information and flexibility, the need to ensure access to technology—this all makes that the transition in trade (and therefore in the supply chain) is starting to pick up even more momentum. Digital consumer is looking not only for his own needs, but is looking also on the impact of his needs. That is why he is putting more and more attention not only to his needs. He is thinking also about his need: how it was carried out in accordance with the natural environment. He wants his need to become stable, and he remembers to protect the environment. Important become concepts of sustainable development and ecological footprint. And with them there is another, new need—to ensure adequate quality of life not only for himself, but also for future generations.

The aim of the article was to reflect on the digital consumer needs. The consumer this is sometimes called a Consumer 4.0. The analogy to contemporary trends of production (Industry 4.0) and logistics (Logistics 4.0) is eyes-whist. Consumer 4.0 uses devices that are connected to the Internet and need to be in constant contact with the world. The article has undergone a short analysis of the needs of the digital consumer and the evolution of these needs over the years refers to the shape of the modern supply chains and networks, in particular their digitization, computerization but also sustainable development.

Particular attention was given to the mutual relations between the needs of the digital consumer and trends in modern supply chains and networks. Therefore who misses important trends in the area of digital delivery network and does not meet the needs of digital consumer, will lose. Appreciation of the needs of the digital consumer and the impact of those needs on the logistics for some will mean growth stimulus, for others can become the beginning of the end.

References

1. Accenture. *Dynamic Digital Customer*. https://www.accenture.com/pl-pl/_acnmedia/PDF-39/Accenture-PoV-Dynamic-Consumers.pdf.
2. Alicke, K., Rexhausen, D., & Seyfert A. *Supply chain 4.0 in consumer goods*. https://www.mckinsey.com/industries/consumer-packaged-goods/our-insights/supply-chain-4-0-in-consumer-goods.
3. Berbeka, J. (2016). Wirtualna i rozszerzona rzeczywistość a zachowania konsumentów [Virtual and augmented reality and consumer behaviour]. *Studia Ekonomiczne/Uniwersytet Ekonomiczny w Katowicach. Ekonomia, 303*(7), 84–101. http://cejsh.icm.edu.pl/cejsh/element/bwmeta1.element.cejsh-1cf50c0e-ca3c-4297-8cad-5c82edbc9983/c/06.pdf.

4. Buxmann, P., & Hess, T. (2008). Software as a service. *Wirtschaftsinformatik, 50*(6), 500–503. https://link.springer.com/content/pdf/10.1007/s11576-008-0095-0.pdf
5. Christopher, M. (2016). *Logistics and supply chain management.* https://books.google.pl/books?id=NIfQCwAAQBAJ&printsec=frontcover&dq=logistics+and+supply+chain+management+christopher&hl=pl&sa=X&ved=0ahUKEwj9yO319MjZAhXRyaQKHdt0CQgQ6AEILzAB#v=onepage&q=logistics%20and%20supply%20chain%20management%20christopher&f=false.
6. Ciszewski, T., & Wojciechowski, J. (2012). Logistyczne zastosowanie systemów informacyjnych [Logistics application of IT systems]. *Logistyka, 3,* 349–352.
7. D'Aveni, R. A. (2010). *Beating the commodity trap: How to maximize your competitive position and increase your pricing power.* Boston: Harvard Business Press.
8. Ericsson. (2018). *10 hot consumer trends.* https://www.ericsson.com/assets/local/networked-society/consumerlab/10hct_report2018_rgb.pdf.
9. Ernst&Young. *The state of sustainable supply chains.* http://www.ey.com/Publication/vwLUAssets/EY-the-state-of-sustainable-supply-chains/$FILE/EY-building-responsible-and-resilient-supply-chains.pdf.
10. Gupta, S., & Palsule-Desa, O. D. (2011). Sustainable supply chain management: Review and research opportunities. *IIMB Management Review, 23,* 236–237.
11. Guttentag, D. A. (2010). Virtual reality: Applications and implications for tourism. *Tourism Management, 31*(5), 637–651.
12. Hyun, M. Y., Lee, S., & Hu, C. (2009). Mobile-mediated virtual experience in tourism: Concept, typology and applications. *Journal of Vacation Marketing, 15*(2), 149–164.
13. Inpost. *Już 5 mln ofert udostępnia dostawę Allegro Inpost* [5 million offers with Allegro InPost delivery], press release. https://magazyn.allegro.pl/4245-juz-5-mln-ofert-udostepnia-dostawe-allegro-inpost.
14. Inpost. *Paczkomaty Inpost zintegrowane z allegro* [Inpost packing machines integretes with allegro e-commerce platform], press release. https://integer.pl/pl/aktualnosci/paczkomaty-inpost-zintegrowane-z-allegro-608.
15. Kawa, A. (2011). *Konfigurowanie łańcucha dostaw: teoria, instrumenty i technologie* [Supply chain configuration: Theory, instruments and technologies]. Poznań: Wydawnictwo Uniwersytetu Ekonomicznego w Poznaniu.
16. Meola, A. *The rise of M-commerce: Mobile shopping stats and trends.* http://www.businessinsider.com/mobile-commerce-shopping-trends-stats-2016-10?IR=T.
17. Money.pl. *P&G: W erze cyfryzacji konsumenci szukają marek zaangażowanych społecznie* [In the era of the digitisation of consumers looking for brands involved socially]. https://www.money.pl/gielda/wiadomosci/artykul/pg-w-erze-cyfryzacji-konsumenci-szukaja,65,0,2290753.html.
18. Patrick, K. *Google's Alphabet relaunches smart glasses for business use.* https://www.supplychaindive.com/news/Google-Glass-Alphabet-corporate-warehouse-logistics/447487/.
19. Schrauf, S., & Berttram, P. *Industry 4.0: How digitization makes the supply chain more efficient, agile, and customer-focused.* https://www.strategyand.pwc.com/reports/industry4.0.
20. Szymczak, M. (2015). *Ewolucja łańcuchów dostaw* [Evolution of supply chains]. Poznań: Wydawnictwo Uniwersytetu Ekonomicznego w Poznaniu.
21. Understanding Digital Consumers. https://www.reachfirst.com/understanding-digital-consumers/.
22. Żuchowski, I., & Brelik, A. Wybrane uwarunkowania wewnętrzne zachowania konsumentów [Selected internal determinants of consumer behavior]. *Zeszyty Naukowe Ostrołęckiego Towarzystwa Naukowego, 21,* 212–231. mazowsze.hist.pl/28/Zeszyty_Naukowe_Ostroleckiego_Towarzystwa_Naukowego/650/2007/23163/.

Value for the Customer in the Logistics Service of E-commerce

Justyna Majchrzak-Lepczyk and Martina Blašková

Abstract Each year sees e-commerce sales records in Poland increasing, which contributes to changes in consumer behavior. Completing orders efficiently plays an increasingly important role, and understanding consumer's needs and their satisfaction becomes the supreme role of e-enterprises. The aim of the study is to present selected aspects of logistic customer service and make its evaluation from the perspective of created values. The source basis of the study is the available subject literature and the author's own research on the perception of logistic customer service of individual e-commerce in Poland. The research was carried out using a questionnaire, posted on the Webankieta.pl platform. 248 feedback questionnaires were obtained. Although they do not have a generalizing character, their illustrative nature determines their practical usefulness. The value is interpreted differently in the literature and is currently an important subject of reflection. Its important meaning that plays a major role in achieving the company's business goals and this is clearly emphasized. Providing customers with value enables economic entities to achieve benefits in terms of both financial and reputational aspects. Thus, by rationalizing processes, entities meet the expectations of their clients.

Keywords Customer logistic service · Value · Customer satisfaction
Innovation

J. Majchrzak-Lepczyk (✉)
Department of International Logistics, Faculty of International Business
and Economics, Poznan University of Economics and Business, Poznan, Poland
e-mail: justyna.majchrzak-lepczyk@ue.poznan.pl

M. Blašková
Department of Managerial Theories, Faculty of Management Science
and Informatics, University of Žilina, Žilina, Slovakia
e-mail: martina.blaskova@fri.uniza.sk

© Springer International Publishing AG, part of Springer Nature 2019
A. Kawa and A. Maryniak (eds.), *SMART Supply Network*, EcoProduction,
https://doi.org/10.1007/978-3-319-91668-2_12

1 Introduction

Customer satisfaction has been one of the most repeated slogans of business recently [11, 17, 19]. This is not without reason, the success of the company is generated by the ability to meet customer needs and expectations by offering them value [2, 7, 18]. A better value for the customer is a generator of success, where the important area to compete is the logistics and marketing system.

The value is one of the most fundamental concepts in management. At the same time in the literature we deal with definitional multiplicity of this concept. Its perception depends on both the expectations of customers as well as the different phases of purchase. Value is therefore a subjective category. The benefits offered by companies should outweigh those of the competition and guarantee satisfaction. Consumers have certain expectations of the expected value of the products and their assessment affects the sense of satisfaction and the likelihood of re-purchase. Thus, the question of value for the customer is an important subject of research, being an essential component of market exchange. Therefore, the aim of the study is to indicate the most important values in the client's opinion, which are created by the logistic service of e-commerce.

2 Elements of Logistics Customer Service

Logistic customer service is increasingly being recognized as a determinant of allowing companies to raise their competitive position. Kempny [17, pp. 24–26; 18, pp. 155–169] defined the logistics customer service, which is primarily reflected the transaction and post-transaction elements, thus indicating that the pre-transaction elements remain in the area of marketing. Each of the activities related to the logistics customer service falls within the scope of three phases.

In the first pre-transaction phase, the key is to determine the detailed characteristics of the components, procedures and regulations in customer service, which are designed to meet all applicable standards. Whether the actions taken are effective depends on the correct diagnosis of the market, properly defined needs and expectations of buyers, which should be taken care of before proceeding with the transaction. Therefore, the first phase involves elements not directly related to logistics, and consists mainly of marketing.

The transaction phase is paid the most attention because of direct contact with the customer. This refers to the flexibility of delivery, convenience of placing orders, availability of product items, execution time, the flow of information, knowledge concerning the orders status and many others. What is important at this stage is therefore constant monitoring and responding to the common undesirable effects. Although overcoming some errors or eliminating mistakes often causes a lot of trouble to companies, it can successfully raise their competitiveness.

Actions related to the post-transaction aspects prolong the contact with the customer. The elements of this phase are responsible for the proper use of products, protection of the interests and health of customers. What is important at this stage is the possibility of further cooperation between the parties in the future. In this respect, one should point to the methods of handling complaints and returns, the system receiving complaints, the duration of the replacement of a defective product, guarantee system, the availability of spare parts, support and technical advice, installation and repairs, tracking the product (important in industries where the product must be immediately withdrawn from the market) [8], or discounts for regular customers.

What is important is the fact that it can not be concluded that there is a clearly defined set of standards of logistic customer service, which could refer to different fields of activity of companies. Customer care will therefore be perceived as the ability, or the skill to meet customer needs particularly in terms of time and place of delivery, by means of a suitable distribution channel, while using available forms of logistics activity, including transportation, warehousing, inventory management, information, packaging, or returns. After determining the most important elements for the client/market segment one should start to measure their performance. Logistics service should ensure customer satisfaction from placing orders to receipt of products. Thus, the service should be seen in main dimensions:

- duration of the contract or the delivery time from the point of view of the customer,
- reliability, which allows to maintain adequate inventory levels and does not require a safety margin in supplies,
- efficiency of communication guaranteeing the transfer of information both within the company and in direct customer-dealer relationship,
- convenience requiring the company's flexibility in terms of all the links in the supply chain.

Therefore, considering the multidimensional approach to customer service, one can assume the areas co-creating logistics customer service. In the literature we find a wide range of measures of the impact assessment and logistics capabilities in the area of resource flows that are classified according to various criteria [10, 18, 21]. Table 1 summarizes the selected metrics, which can contribute to the efficiency of customer logistics in every economic entity, in different phases of the purchase.

Therefore, meters reflect the real state of logistics processes and systems for the movement of products and information of quantity, time, location and quality. Today mere statements of companies associated with the level of service offered to the customer are no longer sufficient. What is indispensable is a systematic program for the measurement and control of service offered to customers. The measurement results should contribute to decision-making on reorganization in a situation where service goals are not implemented properly. The effectiveness of corrective actions performed on the basis of the measurement results can contribute to the creation of an effective program of logistics customer service.

Table 1 Selected measures of the efficiency of customer service [10, 18, 21]

Dimension of logistic customer service	Elements of logistics service	Area of measurement
Time	Delivery time	Cycle length of the contract from the moment of its submission by the customer until receipt of the ordered product
	Certainty	Order processed without damage and defects
Reliability	Completeness	Product's compliance with received orders
	Punctuality	Delivery time compliance with contract provisions
	Availability of product in the supply	Adjusting the assortment to market needs
	Flexibility	The ability of the logistics system to respond to implementation of custom orders (quantity, time, form)
	Complexity	Assortment range
Convenience	Frequency	Restrictions by the supplier relating to the intervals on the execution of orders
	The minimum order	Restrictions by the supplier concerning the minimum allowable contract
	Convenience of getting to the company's office	Location, hours of operation, to transport connection, parking lots
	Staff competences	Advice, product knowledge and their replacements
	Transfer of orders	Convenience of placing orders
Communication	Information on the status of the order	The opportunity to follow/track the implementation of the shipment, the availability of real-time information
	Documentation quality	Clarity and transparency of documentation
	After-sales service	Technical support, service maintenance, complaints, return service efficiency, speed of response to emergency situations

3 The Importance of Supply Networks in Logistics Services

The constantly developing e-commerce market and increasingly demanding clients are one of the biggest challenges of today's enterprises. In e-commerce, the manner and speed of delivery of a parcel play a significant impact on the level of customer satisfaction, however, it is becoming clearer that the effective communication is also vital. The market of logistic services, like e-commerce, is subject to constant changes. It is developing at a very fast pace and expanding the range of offered solutions for the entities it serves. This is mainly due to changes in the management of procurement, production and distribution of goods. E-commerce cannot exist without efficient logistics, and a professional logistics company can be a valuable source of competitiveness, which allows to offer the customer even higher value [16, 31, 32]. This is because deliveries to the end customer constitute the most dynamically changing space in the field of logistics services. Here, the speed and flexibility of adapting to the evolving needs of the customer is essential. For this reason, a number of new solutions are introduced, shortening the delivery time, enabling the development of parcel return systems, evening and Saturday deliveries, the ability to change the delivery address even when the shipment is on the way, extensive personal collection points and more. In view of such a significant role of the client, it is justified to manage integrated, dynamic and flexible supply networks.

Therefore, the changes taking place in logistics meet both e-entrepreneur's and client's needs. Thanks to them, the owners of e-stores have now systems which enable them to prepare a delivery note, order a courier or track shipments. Thus, technologies are revolutionizing the process of sending parcels, improving the functioning of online stores, and shortening the time needed for comprehensive customer service. Logistics of e-commerce, more than in case of the traditional one, requires support from Information Technology. An important role, however, is played not only by specialized software supporting the implementation of logistics processes, but also by an e-mail platform that enables communication with the recipient of the shipment. Therefore, in the organizational and process area, it becomes necessary to partner with suppliers of technological solutions.

All links which build the network have a significant impact on the efficiency and effectiveness of undertaken activities and processes. The rising pressure of the environment which affects modern organizations requires searching for increasingly sophisticated ways to survive and grow. Despite the use of extensive IT systems or technical infrastructure, managing supply networks will always be based on people and interpersonal relations. It is their involvement that will decide how many resources will be used and developed towards integration, so that the entire network can simultaneously react to demand changes.

The digital revolution has significantly influenced the change in the functioning of entities and consumer behavior. Undoubtedly, new technologies have offered greater comfort and speed of obtaining information, thus encouraging the use of online channels not only for communication, but also for commercial purposes. Internet users have become a customer who demands short cycles of services, low costs, transparency and corporate responsibility. Nowadays, independent business operations are difficult in many aspects, which is why more and more companies integrate within the supply network. Their main goal is to strive for comprehensive optimization of material flow. An important element of the supply network is undoubtedly its integrators, acting as a link, that is, the point of contact between more than one supply chain. Thus, the coordination of individual activities, integrating the resources of individual participants, undoubtedly leads to completing a specific task at a high level. Online sales requires from e-entrepreneurs the organization of complex logistic processes, in particular: storage, packaging, shipping, handling returns and complaints, or post-trade service. Thus, the combination of all resources, including warehouse and IT facilities, allows to create an offer which helps store owners to focus on their key competences and sell their products, while logistics processes are moved to companies—players from the network. Therefore, apart from a high standard of offered services, you can complete many orders, at the same time shortening the time of order fulfillment and fully satisfying the final recipient. This, in turn, often offers them additional value.

4 Shaping Value for Customer

Building the company's market position can take on different dimensions. One of them, which is the basis of considerations, is based on the concept of creating value for the customer.

"Value" can be defined in different ways, depending on the scientific discipline in which it will be considered. For the purposes of this article it is assumed that in the global market—value should be considered for its ability to create solutions, benefits, satisfaction and emotional experience in the minds of consumers [5, pp. 30–34]. Thus, value for the customer determines the need to analyze trends and market developments which shape the needs and customer behaviors [17]. Thus, the subject concept is not easy to define because of subjectivity and its different perception on the part of various market players. Its three-dimensional potential was shown in Fig. 1.

What plays the important role is communication between the different parties (suppliers, manufacturers, contractors, etc.). It allows exchange of information about the expected value (products and services). In Polish and foreign literature the notion of value is linked to the concept of marketing management. Most commonly

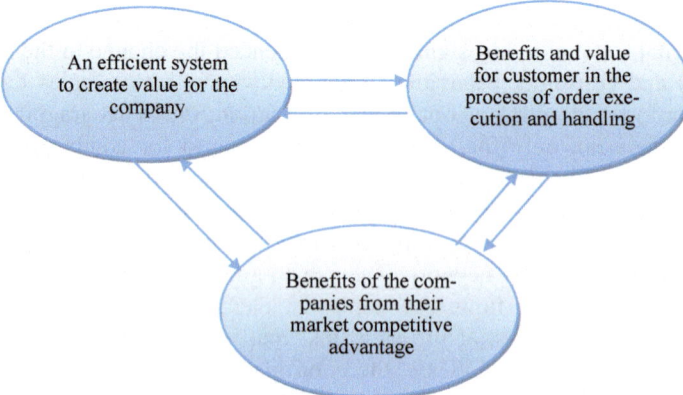

Fig. 1 The three-dimensional strategic potential

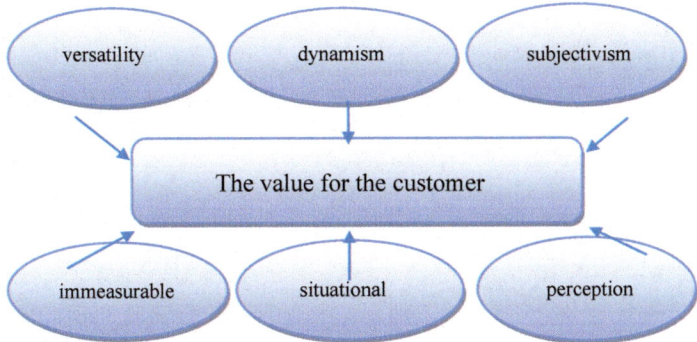

Fig. 2 The basic attributes of value for customer [15, pp. 76–77]

the term value for the customer means how attractive the offer is according to the customer.

Speaking of value for the customer one needs to be aware of the distinction between the expected value and the finally obtained value. Szymura-Tyc [27, p. 75] separately defines the expected value, which is "additional, subjectively perceived expected customer benefits and costs associated with the acquisition and use of the product" and the value obtained: "additional, subjectively perceived received by the customer benefits and costs associated with the acquisition and use of the product".

It is clearly visible that the category of "value" is characterized by high complexity, which boils down to differences in interpretation of the term. Seeking common features, it is clear that the value for the customer is associated most often with the relationship between the perceived customer benefits and costs that they must cover.

However, it may take a different dimension depending on the phase of purchasing, i.e. before the purchase, during the purchase and after the purchase [17].

In the first phase of purchasing the client focuses their attention on the attributes of the product and the search for possible alternatives. The other two phases are closely connected with the feelings associated with the use of the product. The most commonly considered in the literature basic attributes of value for the customer are presented in Fig. 2.

The first feature is the subjectivity of value for the customer, which is determined by individual needs and preferences. The situational nature of value is the relationship between costs and benefits in a specific situation in which the product was purchased/used. Another feature is the perceived value, i.e. assessment of the value received by the customer, taking into account only those costs and benefits which are perceived. Dynamism, as another attribute, is characterized by high volatility of the product while using it. This volatility, therefore, is dependent on the cycle of life and the benefits and costs occurring at that time. Versatility and immeasurability as important attributes are applicable to every buyer and represent important features which should be tested.

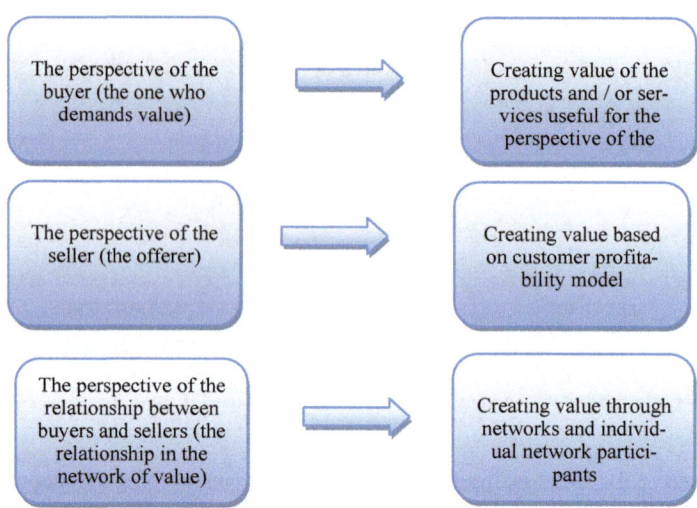

Fig. 3 Perspectives for the value creation process [27, pp. 6–7]

When considering the value creation process, one should pay attention to the fact that value can be seen in many different aspects by various participants in the market exchange (Fig. 3). The value associated with the activities of economic entities can be regarded as a kind of attraction, by means of which customers evaluate the offered products.

The multitude of products and their similarity both visual and technical means that the customer is not able to see significant differences between the offers. This phenomenon raises the need to differentiate from the competition through additional advantages and value. It is therefore necessary to put emphasis on professional service offered.

Problems of customer service are within both marketing and logistics. Marketing through the use of its tools and analysis allows the identification of needs and future demand, creating new products [4, 13]. It makes certain promises to customers at the same time. The fulfillment of these promises and satisfying the demand is carried out by logistic in managing transport and storage processes. Thus, in order to properly maximize in meeting customer needs while minimizing costs, it is necessary to perceive the mechanisms and relationships that occur between marketing and logistics. Therefore, a lot of activities in the enterprise shapes the value for the customer, each resulting and depending from each other (Fig. 4). Management value category takes the dimension of both operational and strategic nature.

Shaping the relationship with the customer, so important in the modern world, should not be considered as a task only for marketing, because logistics has a very large share in the process. Whether the delivery will be completed on time, whether the goods will land on store shelves at the right time in the right amount—it all undoubtedly contributes to customer satisfaction. Thus, the value of the product is

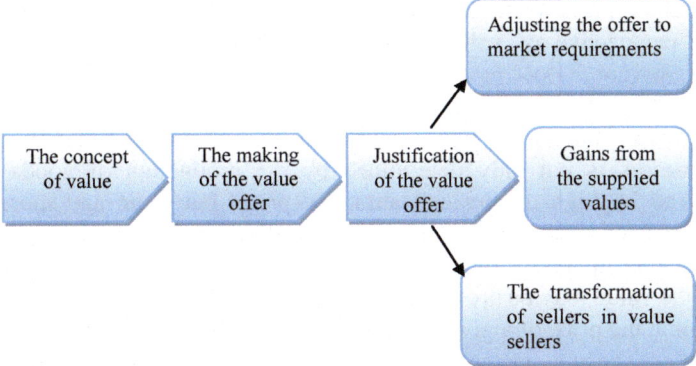

Fig. 4 Managing customer value by Anderson et al. [1, p. 24]

the perception of the whole offer, understood as the product and accompanying services, which is most evident in e-commerce.

Increasing customer awareness requires companies to look for new sources of competition, since many of the measures used so far have lost their importance. In this context, logistics customer service is an important area of raising customer satisfaction, allowing to strengthen the competitive position of economic agents.

Logistics customer service is particularly important in an era of intensively flourishing commerce. The characteristic features of electronic commerce, among others, are: lack of direct verbal and non-verbal contact with the seller, lack of contact with the product or the need to deliver goods to a large number of individual recipients. These features are also major differences in relation to traditional trade and are determinants in making purchase decisions. To offset the negative effects of these differences for customers, the level of logistics customer service offered by online stores, which are the basic units of electronic sales, should be as high as possible. In electronic commerce the critical elements are: transactional elements of logistics services, especially the availability of products and services, the speed of communication between the partners of commercial operations, execution time, range of activities, flexibility and reliability of deliveries.

Customer service creating logistical value should include activities which allow to meet the needs of widely accepted standards and innovative activities which go beyond the standards and form the uniqueness of the offer.

5   Creating Value in Logistics—Research Results

The e-commerce market in Poland is systematically growing, and from the Gemius report [6] "E-commerce in Poland 2016" it appears that almost half of Internet users make e-purchases. Poland is this European country where the e-commerce market is developing the fastest, and online shopping is becoming something natural for

Poles. Therefore, it is worth discussing the importance of logistics customer service from the perspective of values it creates with reference to the popularity of purchases.

Undoubtedly, creating value for the customer in the supply chain is a more difficult process than in a single enterprise, due to the complexity. Therefore, competitiveness is not only determined by the production of goods, which is characterized by high quality and attractive price, but more and more often by customer service which guarantees smooth flow of information and financial resources along the entire supply chain [5].

Thus, the value may be provided by aggregate benefits, allowing the customer to conclude that the price of a given product is reasonable, even if other entities have a cheaper offer. The activities which can to meet the expectations of customers are the ones which determine the company's success [22].

The research was guided by the goal of identifying how customers in Poland perceive the logistic service of online stores and what value means for them in this aspect.

The research was quantitative and carried out using an online questionnaire, based on a questionnaire in the electronic version, posted on the Webankieta.pl platform. 248 feedback surveys were obtained, and these were analyzed. The study involved adults who make purchases in e-commerce. The selection was random, which means that the study is not generalizing.

The majority of respondents were women—67%, and young people up to 29 years of age, with a nearly 59% share in the study. Almost 60% of clients had a university degree, 40% had a secondary education. Respondents most often declared being a student—over 38%, while also often being professionally active. Available income per one person was up to PLN 2000 for almost 35% of respondents, 29% of them declared salaries up to PLN 3000. Among the rest of respondents, there was a large variation in earnings. 42% of people lived in large urban centers, a big group of clients was characterized by large dispersion due to their place of residence (Fig. 5).

Over 66% of people participating in the survey have been making purchases online for more than 4 years. When asked about the elements which in their opinion affect the perception of the website/online store as a reliable one, respondents most often indicated the need for obtaining detailed information related to the terms of how order or possible refund or complaint are made (80%). The price of the product itself and the transparency and functionality of the website are also very important (nearly 60% of indications). Modern consumers expect the product range to be personalized in terms of aesthetic values and individual interests [30]. There is a growing demand for non-standard goods and services tailored to the tastes and preferences of market participants. So if the product does not meet their expectations, taking advantage of the right of return becomes an extremely important issue—52% of clients have used this right.

Therefore, for a modern e-client, proper communication on the shop-client line is of great value [20, 28]. As the accuracy and legibility of information about the

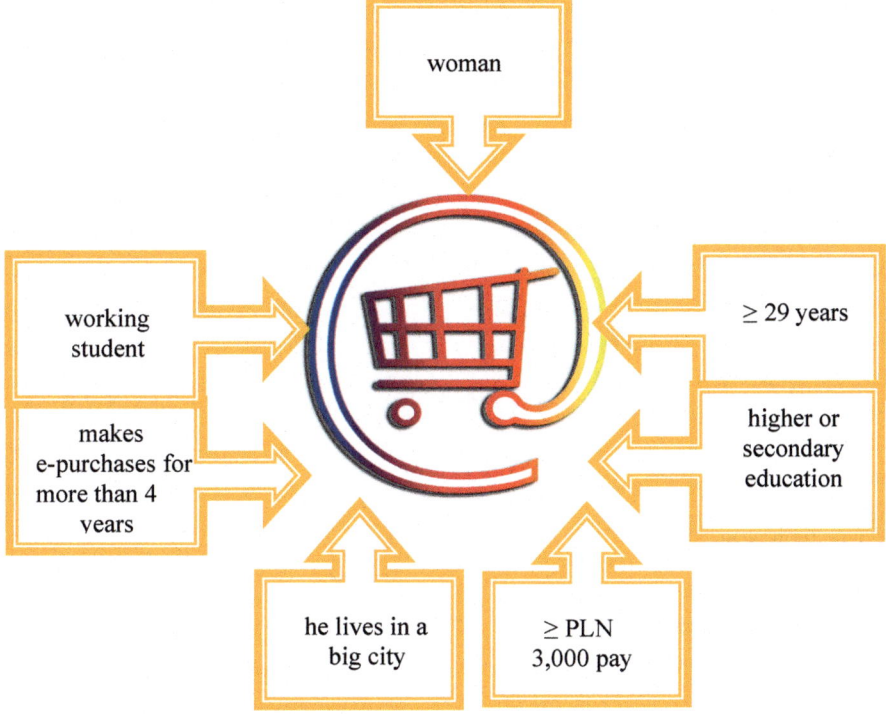

Fig. 5 Average e-consumer (research referent)

terms of the contract, complaints or refund is usually important, 96% of respondents also took notice of the functionality of the website.

This leads to the ease of placing orders (89%), which is additionally encouraged by the clear and good quality presentation of products (88%) and a wide range of products (85%). In the next stage of the purchasing process, prices and the possibility to choose a convenient form of payment [14, 24] (83% of indications) become important (Fig. 6).

Relationships [19, 30] and interpersonal contacts, shaped, among other things, by product [25], recommendations, or posted opinions on forums, portals or elsewhere about the company and the products it offers, also shape credibility and lead to a desire to make a purchase (80% people taking part in the study). In Fig. 7, other factors have been presented which were indicated by the clients as valuable.

The importance of responding to the reported problem and previously indicated assortment availability (84%) are important for the professionalism of customer service for the vast majority of people. Shopping security is equally important.

The process of order fulfillment, as well as possible complaints or return of goods (81% each) affect the opinion on logistics service. In addition, a good contact with the seller, who informs about the status of the order and provides answers to

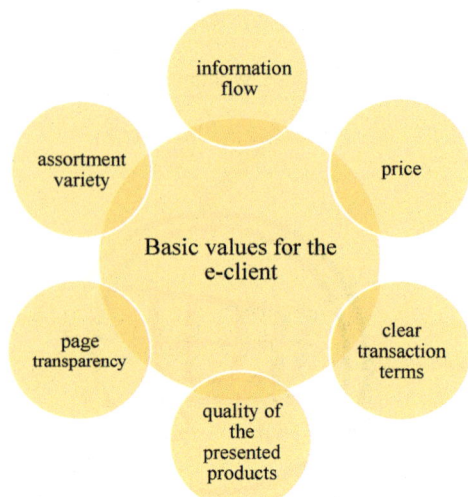

Fig. 6 Basic values for the e-client

Fig. 7 Perception of the credibility of the website/online store

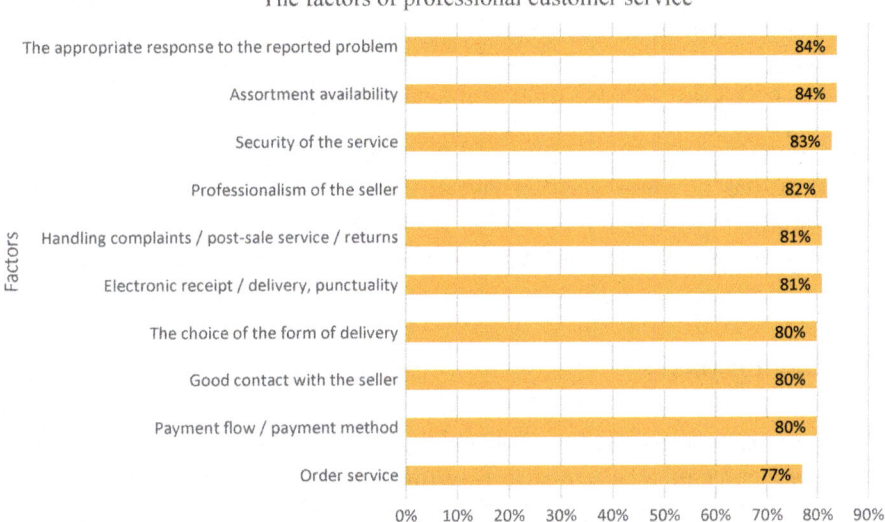

Fig. 8 The factors of professional customer service

questions asked by the customer, can undoubtedly influence the perception of the store as a professional one (Fig. 8).

In addition to the positive aspects of service, customers also indicated factors limiting its efficiency.

All clients indicated poor organization of the work of a given entity as the main reason for the lack of professionalism. For 57% of respondents, the lack of proper information exchange between the buyer and the seller plays an important role, which has been indicated earlier. Therefore, if there is no information flow, such an entity is not perceived as a professional one. In addition, if the company is sluggish in completing the order, which contributes to a prolonged deadline for the contract, lack of availability or delay in delivery, this kind of seller is wrongly perceived. As you can see, there are many areas which, in the customer's opinion, shape logistic service and affect its professionalism.

Customers increasingly use mobile solutions that allow them at any time and place to familiarize with the offer of enterprises, compare the attractiveness of individual products, view opinions and comments, place orders or make payments [26]. The authors of the report They Say They Want a Revolution are also paying attention to the growing need to use mobile solutions in the purchasing process. Total Retail 2016 [25]. Mobility is therefore one of the features determining the development of supply chain management in e-commerce using mobile technologies. The use of mobile applications and devices allows you to achieve the desired effects, such as reducing costs or increasing the ability to respond to changes, resulting in a competitive advantage [9]. Zhu et al. [31] perceive the development of mobile technologies as one of the main trends in the digital management of the supply chain.

The conducted study allowed to draw many interesting conclusions, which indicated the areas that create value in logistic services among Polish e-clients [6, 29]. These are undoubtedly: a broadly understood flow of information, because it concerns both the store's offer and communication with the client, or the response to the reported problem. The next, most often indicated values are: assortment of the given store and opinions and recommendations from other people. In conclusion —lack of direct contact between the e-commerce client and the seller must be compensated by offering professional communication which surely shapes and influences the quality of logistics services.

6 Conclusions

Thus, the value may consist of aggregated benefits, allowing the customer to conclude that the price of a product is justified, even if others have a cheaper offer. The efficiency of a business is determined by the actions that are able to meet the needs of customers [13]. It has become necessary to seek innovative opportunities to achieve success in the market, offering the user a better experience than the previously existing solutions. The nature of trading and logistic activities is clearly changing, thus determining the need for constant versification of the existing business models. Observing the ongoing transformation of the market, one should expect to find the need for fusion of electronic and traditional forms of sales. E-commerce is meeting the growing demands of the consumer, creating new trends and market behaviors. Without a doubt, while analyzing consumer preferences, companies using ICT technologies, are growing in strength and gaining a competitive advantage [23].

Summing up, it is reasonable to build and maintain advantageous relationships and network connections with other market players, which will undoubtedly allow to develop common standards and enable the use of previously unreachable sources of competitiveness. Conducted research did not include the strategic potential of enterprises and their cooperation within the configuration of supply networks, but only the effects they achieve in the opinion of customers. In this respect, resources and competences are often not perceived by customers at all, but they are an important determinant of market success.

Due to the diversity of products and rapidly changing expectations and buyer preferences customer service cannot be seen as a series of repeated modules, but rather as a system of connected vessels, the essence of which is the best service offered to end customers [3]. The companies which cooperate with other entities in the supply chain, are forced to see customer service as a more complex system in which the participants are all cooperating companies [2, 3, 7].

The article presents the essence of logistic customer service of e-commerce in value creation. It is evident that service plays an increasingly important role, and its quality translates into the profitability and competitive position of economic entities.

The obtained results were used to identify factors which increase the quality of logistics services and to define the main attributes of the service logistics value in the customer's opinion. Market success results from the ability of fast economic agents to respond to the variability of the environment. Mobile technologies are a kind of catalyst accelerating the development of the e-commerce supply network, perfecting business processes and revitalizing the possibilities of responding to the needs of modern consumers. Thus, the modern e-business model requires a flexible and adaptive design, to create and deliver value to increasingly demanding customers.

It is necessary to have knowledge about customer behavior, their preferences and expectations for products. Thus, providing value requires the ability to transform company resources into product usability collections, which are a source of value for customers [12, p. 109].

The modern consumer has more and more choice options, thanks to the modern technological tools which allow both to obtain interesting content and make purchases using them. The speed at which innovative solutions are introduced onto the market triggers the necessity of constant observation and conducting further, in-depth research in this field.

References

1. Anderson, J. C., Kumar, N., & Narus, J. A. (2010). *Sprzedawcy wartości* [Value sellers]. Warszawa: Oficyna a Wolters Kluwer Business.
2. Bellingkrodt, S., & Wallenburg, C. M. (2015). The role of customer relations for innovativeness and customer satisfaction: A comparision of service industries. *The International Journal of Logistics Management, 26*(2), 254–274.
3. Cahill, D. L., Goldsby, T. J., Knemeyer, A. M., & Wallenburg, C. M. (2010). Customer loyalty in logistics outsourcing relationships: An examination of the moderating effects of conflict frequency. *Journal of Business Logistics, 31*(2), 253–278.
4. Chiu, Ch. M, Wang, E. T. G., Fang, Y. H., & Huang, H. Y. (2014). Understanding customer's repeat purchase intentions in B2C e-commerce: The roles of utilitarian value, hedonic value and perceived risk. *Information Systems Journal, 24*, 85–114.
5. Christopher, M. (2016). *Logistics and supply chain management* (5th ed.). UK: F.T Publishing.
6. Chuang, S. H., & Lin, H. N. (2015). Co-creating e-service innovations: Theory, practice, and impact on firm performance. *International Journal of Information Management, 35*, 277–291.
7. Cichosz, M., Goldsby, T. J., Knemeyer, A. M., & Taylor, D. F. (2017). Innovation in logistisc outsourcing relationship—In the search of customer satisfaction. *LogForum, 13*(2), 209–219.
8. Ciesielski, M. (2011). *Zarządzanie łańcuchami dostaw* [Supply chain management]. Warszawa: Wydawnictwo PWE.
9. Eng, T. Y. (2006). Mobile supply chain management: Challenges for implementation. *Technovation*, no. 26, s. 682.
10. Gao, T. T., Rohm, A. J., Sultan, F., & Pagani, M. (2013). Consumers un-tethered: A three-market empirical study of consumer's mobile marketing acceptance. *Journal of Business Research, 66*, 2536–2544.
11. Garbarski, L. (red.). (2011). *Marketing. Koncepcja skutecznych działań* [Marketing. The concept of effective actions]. Warszawa: Wydawnictwo PWE.

12. Gemius. (2016). *E-commerce w Polsce 2016* [E-commerce in Poland 2016]. https:// ecommercepolska.pl/files/9414/6718/9485/E-commerce_w_polsce_2016.pdf. December 1, 2016.
13. Geng, X., & Chu, X. (2012). A new importance-performance analysis approach for customer satisfaction evaluation supporting PSS design. *Expert Systems with Applications, 39*(1), 1492–1502.
14. Heiskanen, P. (2016). *E-Commerce payment methods-From traditional to online store*. Karelia Ammattikorkeakoulu.
15. Jiang, L., Jun, M., & Yang, Z. (2016). Customer-perceived value and loyalty: How do key service quality dimensions matter in the context of B2C e-commerce? *Service Business, 10*(2), 301–317.
16. Kawa, A., Maryniak, A. (2018). Lean and agile supply chains of e-commerce in terms of customer value creation. In A. Sieminski, A. Kozierkiewicz, M. Nunez, & Q. T. Ha (Eds.), *Modern approaches for intelligent information and database systems* (pp. 1–11). Switzerland: Springer International Publishing AG.
17. Kempny, D. (2001). *Logistyczna obsługa klienta* [Logistic customer service]. Warszawa: Wydawnictwo PWE.
18. Kempny, D. (2008). *Obsługa logistyczna* [Logistic service]. Katowice: Wydawnictwo AE w Katowicach.
19. Kim, Y., & Peterson, R. A. (2017). A meta-analysis of online trust relationships in e-commerce. *Journal of Interactive Marketing, 38*, 44–54.
20. Kościelniak, H. (2014). An improvement of information processes in enterprises—The analysis of sales profitability in the manufacturing company using ERP systems. *Polish Journal of Management Studies, 10*(2), 65–72.
21. Kramarz, M. (2014). *Elementy logistyczne obsługi klienta w sieciach dystrybucji. Pomiar, ocena, strategie* [Logistic elements of customer service in distribution networks. Measurement, evaluation, strategies]. Warszawa: Wydawnictwo Difin.
22. Lan, S., Zhang, H., Zhong, R. Y., & Huang, G. Q. (2016). A customer satisfaction evaluation model for logistics services using fuzzy analytic hierarchy proces. *Industrial Management & Data Systems, 116*(5), 1024–1042.
23. Mentzer, J. T., Myers, M. B., & Cheung, M. S. (2004). Global market segmentation for logistics services. *Industrial Marketing Management, 33*, 15–20.
24. Nica, E., & Potcovaru, A. M. (2015). Gender-typical responses to stress and illness. *Psychosociological Issues in Human Resource Management, 3*(2), 65–70.
25. PwC. (2016). *They say they want a revolution*. Total Retail 2016.
26. Ström, R., Vendel, M., & Bredican, J. (2014). Mobile marketing: A literaturę review on its value for consumers and retailers. *Journal of Retailing and Consumer Services, 21*, 1001–1012.
27. Szymura-Tyc, M. (2005). *Marketing we współczesnych procesach tworzenia wartości dla klienta i przedsiębiorstwa* [Marketing in contemporary processes of creating value for the client and enterprise]. Katowice: Wydawnictwo Akademii Ekonomicznej.
28. Wu, Y. L., & Li, E. Y. (2018). Marketing mix, customer value, and customer loyalty in social commerce: A stimulus-organism-response perspective. *Internet Research, 28*(1), 74–104.
29. Wynstra, F., Spring, M., & Schoenherr, T. (2015). Service triads: A research agenda for buyer-supplier-customer triads in business services. *Journal of Operations Management, 35*, 1–20.
30. Yadav, M. K., Rai, A. K., & Srivastava, M. (2014). Exploring the three-path mediation model: A study of customer perceived value, customer satisfaction service quality and behavioral intention relationship. *International Journal of Customer Relationship Marketing and Management (IJCRMM), 5*(2).

31. Zhu, X., Song, B., Ni, Y., Ren, Y., & Li, R. (2016). *Business trends in the digital era. Evolution of theories and applications* (pp. 63–82). Singapur: Springer & Shanghai Jiao Tong University Press.
32. Zou, H. M., Fang, Y., McCole, P., Qureshi, I., Sun, H., & Ramsey, E. (2016). *Investigating the nonlinear and conditional effects of trust on effective customer retention—The role of institutional contexts*. Australasian Conference on Information Systems, Wollongong.

The Influence of Prosumers on the Creation and the Process of Intelligent Products Flow

Kazimierz Cyran and Sławomir Dybka

Abstract A growing number of intelligent products available in the market makes them pay attention. The areas of use of intelligent products are wide, they can include production processes, supply chains, asset management and product life cycle management as well as direct products offered to final customers. In production activities, intelligent products are to streamline production planning and control, allow products to be tailored to individual needs, and make the choice between the variants made more effective. In intelligent supply chains it is possible to improve the efficiency of sending and receiving goods, improving the control and safety of loads, and implementing environmentally-friendly rationalization. Final customers expect more intelligent comfort, ease of use, higher level of environmental performance, time savings, higher efficiency and performance, safe use and reliability from intelligent products. However, despite many potential benefits offered to users, intelligent products are a major challenge for manufacturers. The aim of the article was to present the role that prosumers play in the process of creating and implementing intelligent products and to indicate the conditions of this process. For this purpose the basic assumptions of the concept of intelligent products and the phenomenon of prosumption are presented. The model of the process of creating intelligent products was also proposed, taking into account the interaction between prosumers and enterprises. The attention is paid to the specificity and contribution of startups to the process of generating innovations, including intelligent products. In the final part, the thread of negative consequences of prosumption and the implementation of intelligent products, which are less frequently discussed in the literature, was featured.

Keywords Intelligent products · Prosumer · Creation · Management

K. Cyran · S. Dybka (✉)
Department of Marketing and Entrepreneurship, University of Rzeszów,
Rzeszów, Poland
e-mail: slawekd@ur.edu.pl

K. Cyran
e-mail: kcyran@ur.edu.pl

© Springer International Publishing AG, part of Springer Nature 2019   241
A. Kawa and A. Maryniak (eds.), *SMART Supply Network*, EcoProduction,
https://doi.org/10.1007/978-3-319-91668-2_13

1 Introduction

An important problem for modern enterprises operating in a competitive market may be the introduction to an offer of intelligent products which are a manifestation of a high level of innovation. Increasingly more often it is noticed that prosumers become the inspiration for work on new solutions, i.e. consumers who, based on their knowledge and experience, suggest modifications of products in the direction of increasing their functionality or providing new values. Authors recognize that the phenomenon of prosumption is discussed in the literature, but it rarely refers to the design and implementation of intelligent products. The literature focuses on the advantages of prosumption and intelligent products, neglecting the negative effects of implementing smart products or the risks associated with the development of prosumption. Thus the article strives to demonstrate relationships between prosumers and the implementation of intelligent products. In this context the aim of the article is to present the role of prosumers in the process of creating and implementing intelligent products and to indicate the conditions of this process. The source of the deliberations was the scientific literature addressing the problems discussed in the article, as well as the authors' own experience.

2 Intelligent Products as a Component of the Smart Supply Network—Conceptual Framework

The smart supply network means greater efficiency and reliability. In contrast to the existing models and solutions, smart optimization means a significant increase in automated and digital solutions. SSN stands for implementations of more and more intelligent solutions, independent, automatic and making the optimal choice, able to react to changes in the environment. In B2B relations, it means introducing intelligent products on the market that essentially have the same assumptions as on the B2C market. Therefore, products and intelligent solutions should be considered as determining the directions of development of products offered to final consumers but also entities in the supply network.

In the digital revolution has changed the conditions and ways of functioning of the economy and contributed to the emergence of new forms of organization and production such as co-configuration or knotworking, based on the cooperation of dispersed producers and consumers within virtual networks to create intelligent products and services. An ability of the organization to create knowledge and learn in the virtual space within the value chain has become the key condition for success [35]. The necessity of constant product development requires constant initiation of learning processes based on the interaction between users, manufacturers and service at both the unitary and organizational levels [6]. Intelligent enterprises understand the importance of knowledge and treat it as the strategic capital of the enterprise. They are focused on acquiring knowledge related to innovations, which allows for a better response to the challenges of the current economy based on knowledge. Intelligent organizations adapt more easily to the changing environment, they are more competitive and resilient to market failure [17].

The products are becoming more and more intelligent, and even interactive, which creates more opportunities to apply new technologies in a new way. It can be seen that few changes push the Industry 4.0 concept to the next level. First, intelligent digital products that can be developed by companies as combined and self-aware products that can provide information on consumption, usage level and storage conditions, data on health status, location and even the customer's mood. Secondly, emerging new technologies that reach a turning point in terms of readiness and availability. Some of them include cognitive calculations, 3D printing, augmented reality, blockchain, machine learning, artificial intelligence, user-controlled voice interfaces and robotics. Thirdly, changing customer attitudes, behaviors and expectations, positioned the Internet of things to give businesses and organizations even more opportunities to collect data and use them to improve customer results [14].

The goals for Intelligent Products are different and depend on manufacturing, supply chains, asset management and product lifecycle management. In manufacturing these products are to improve production planning and control, to enable customized products and to make change-over between variants more effective. In the supply chain intelligent product allow to improve the efficiency of sending and receiving goods or to improve the security of the supply chain. By maintaining the identity of the product or shipment it is possible to pinpoint where thefts occur or to verify the authenticity of the item and reduce the risk of forgery. Another goal is to improve of the rerouting of products and shipments in transit. Especially for the delivery of components for complex systems where delivery to the customer site is time critical. Expensive assets such as tools and equipment often-need to be used by many parties, and their continued use require the services of different service providers. By introducing intelligence to the assets it becomes easier to share assets and also to service them. An applications of Intelligent Products can be extended across many stages in the product lifecycle, an application initially developed for controlling customization in manufacturing may be used for improved handling in the supply chain, and to support efficient maintenance in asset management [23].

Every day customers notice such an examples of products like the intelligent control a growing array of home devices, transmitting data about their use back to manufacturers. Intelligent, networked industrial machines autonomously coordinate and optimize work. Products stream data about their operation, location, and environment to their makers and receive software upgrades that enhance their performance or head off problems before they occur. Products continue evolving long after entering service. The relationship between products and customers is becoming continuous. Intelligent, connected products require a whole new supporting technology infrastructure. Complex product operations can be controlled by the users through numerous remote-access options. That gives users the unprecedented ability to customize the function, performance, and interface of products and to operate them in hazardous or hard-to-reach environments. The combination of monitoring data and remote-control capability creates new opportunities for optimization. Algorithms can substantially improve product performance, utilization, and uptime, and how products work with related products in broader systems, such as intelligent buildings and intelligent farms. Thus the combination of monitoring data, remote control, and optimization algorithms allows autonomy. Products can learn, adapt to the environment and to user preferences, service themselves, and operate on their own [27].

As [21] defines Intelligent Product consists of five fundamentals proprieties: possesses a unique identification, a capability of communicating effectively with its environment, an ability of retain or store data about itself, deploys a language to display its features, production requirements, and a capability of participating in or making decisions relevant to its own destiny [8]. The intelligent product may be considered with 3 levels of intelligence. When the Intelligent Product is only capable of "information handling", it is not in control of its own life, as full control of the product is outside the product. "Problem notification" means that more Intelligent Product can notify its owner when there is a problem but the product is not in control of its own life, it's only able to report its status. Finally the "decision making" level when product can completely manage its own life, and is able to make all decisions without any external action. However, the location of intelligence can be described as intelligence through network when product is completely outside the physical product—the product only contains a device that is used as an interface to the intelligence. Whereas in Intelligence at object all the intelligence, whether this is only information handling, or advanced decision making, takes place at the physical product itself. And finally the aggregation level of the intelligence consist of Intelligent item where the object only manages information, notifications and decisions about itself or Intelligent container which not only manages data or decisions about itself, it is also aware of the components that it is made of and may act as a proxy device for them. If the intelligent container is disassembled or parts are removed or replaced, the parts may be able to continue as intelligent items or containers by themselves as the domain of supply chain management is an intelligent shelve, which can notify its owner when a specific product is out of stock [23].

In the known conceptualization, there are considered the dimensions as distinct from each other and view product intelligence as a multidimensional, construct that is formed by the six described dimensions. The first intelligence dimension—autonomy—refers to the extent to which a product is able to operate in an independent and goal-directed way without interference of the user. An ability to learn is the second dimension and refers to a product ability to improve the match between its functioning and its environment. Reactivity is the third dimension of intelligence and refers to the ability of a product to react to changes in its environment in a stimulus/response manner. Reactivity can be distinguished from the ability to learn in that reactivity refers to instant reactions to the environment. The fourth dimension of product intelligence is the ability to cooperate with other devices to achieve a common goal. Products are becoming more and more like modules with in-built assumptions of their relationships with both users and other products. An increasing number of products are, thus able to communicate not only with their users but also among themselves. The fifth dimension, human-like interaction, concerns the degree to which the product communicates or interacts with the user in a human way. Intelligent products are sometimes able to communicate with their users through voice production and recognition. The last dimension, personality refers to an intelligent product ability to show the properties of a credible character. Physical products can also be equipped with an interface that shows personality characteristics. Even with the ability to show emotions [29].

Intelligent Products were first discussed by [15] in an after sales and service context in 1988 with computers running programs that tracked the configuration and performance, and that could request for service and maintenance. Later the idea of integrating intelligence and control into the product spread to manufacturing [22] and supply chain control [16]. When product individuals in a logistic/production setting are given a traceable individuality, associated content (delivery terms, contract terms, exceptions), and also decision is delegated, the realm of Intelligent Products starts. Such Intelligent Products will have the means to communicate between themselves and also with logistic service providers. Intelligent Products link the Auto-ID technology to the agent paradigm and Artificial Intelligence. Intelligent Products can also play an essential role in product lifecycle management by their capability of collecting usage information and reacting, like estimating needs for repair. By using sensor technologies like thermal, acoustic, visual, infrared, magnetic seismic or radar the conditions of products can be continuously monitored. The access to information on how products have been used could significantly improve the way that products are recycled when they arrive to their end-of-life. Sensor technologies can also contribute to improvements in manufacturing nodes and to the logistics of the entire supply chain, by giving real-time identification, location and other conditions of the products [15].

There can be indicated some of the challenges associated with the design and structure of the intelligent product environment. Modeling Approaches assume that identification of modeling environments should support representation of the product, its information and decision environments, representation of suitable the industrial environment where the product operates and representation of the existing planning and control systems. In the area of the language, product intelligence implies a common language for all related orders and resources or a means of interpreting between tasks required by orders and functionalities supplied by resources. In the sphere of interaction architecture development of system architectures should support the interaction between multiple orders/products and multiple resources across multiple organizations [21]. Creation of original and modern system products is usually done by generating new knowledge of stakeholders involved in the process, among which the "final link" is the most important—the target recipients, with a bundle of reported needs and preferences, consumption patterns. Product life cycles are shortening, and increasingly demanding customers expect market offers that will be able to perform various functions implemented in the core/several cores of a given good or service [1]. In this context, the implementation of intelligent products fits into the user experience (UX) concept at both stages phase of the project as well as production and during the subsequent stages of the product life cycle [36]. UX is a widely spread concept in the field interaction design and related area, and its importance for the successful development of interactive systems, products and services cannot be overestimated. An advanced technological development has become in several senses an integrated part of human daily life. UX is defined [34] as the totality of the effect or effects felt by a user as a result of integration with and the usage context of a system, a device, or a product including the influence of usability, usefulness, and emotional impact during interaction and savoring memory after interaction.

3 The Idea of Presumption in the Context of the Smart Supply Network

The concept of smart supply network is related to the implementation of smart products. It should be understood that the recipients of these technologies may also be the source of innovative solutions. These clients provide information about their needs, suggest improvements. Their knowledge of experience and, above all, commitment and willingness to join the process of improvement, cause that they become prosumers. It is necessary to associate smart technologies and intelligent products as well as prosumers with SSN. Without information collected from intelligent products and prosumers development of smart solutions in supply chain would not be possible.

The term "prosumption" comes from the combination of the words production and consumption and means the interweaving of consumption and production processes, which leads to the blurring of the boundaries between them, as a result of this process consumers become producers at the same time [30]. Despite the fact that prosumption is not a new term and takes up a lot of space in the literature, it is still not clearly defined. The fundamental differences in interpretation result from the evolution of the phenomenon, which is a consequence of changes in the producer-consumer relationship. A precursor that introduced prosumption in literature in the 1980s was A. Toffler. He considered the development of humanity in the context of three consecutive technological waves. The first stage (pre-industrial), initiated by the agrarian revolution, was characterized by the lack of division into production and consumption, as well as suppliers and recipients of goods. The second wave is industrialization, which has brought a separation between passive consumers and profit-making producers. On the other hand, the third post-industrial wave, which is distinguished by the IT revolution, messes with the first stage, gradually blurring the functions of production and consumption. Toffler identified the third wave directly with the emergence of new technologies enabling unrestricted communication between individuals, thanks to the development of services that depart from mass production. At the same time he pointed to the disturbing phenomenon of consumption growth and the possibility of its limitation thanks to prosumption. Toffler treated a prosumer as a person who is willing to take over some of the tasks previously performed by the manufacturer [33].

Today prosumption is understood much more broadly. It is referred to as intelligent consumption and falls into the consumer trends. On the one hand, it appears more and more often in contemporary literature, and on the other hand, it includes an increasingly wider collection of various forms of buyers' activity. There are many similar terms in the literature, such as: partner production, co-creation of values with clients, consumer engagement in a virtual environment [11] and Web 2.0 [7], open innovation [4], collective intelligence [20].

According to [29], prosperity means redefining the role of the consumer, who, by searching for new ways of using existing products, contributes to creating their more perfect versions, or creating absolute innovations. Wolny [38] and [39] claims that prosumption is simply a new way of achieving the goal of increasing the company's profit, and [24] in turn indicates it as a way to reduce the cost of buying and using products by the consumer.

Currently, as Toffler points out, a prosperous lifestyle where individuals become producers, design or modify products according to their own preferences is shaped. According to [32], a new consumption model is created where customers actively and continuously participate in the creation of products, and thus—the importance of prosumption changes as well, which is reflected in companies losing control over products, providing customers with the right tools and materials, partnership and sharing results (users want to take part in benefits, want their commitment to pay off).

The development of prosumption is also connected with activities aimed at environmental protection. An example of this is the growing interest in acquiring energy from renewable sources. The causative factor here is not the economic viability of, for example, own energy production, but the responsibility for the environment. The energy produced by consumers—producers do not damage the environment as much as the corporations do [9]. Prosumption can, therefore, lead to rationalization and limitation of the use of some services, and can be used for practical implementation of the concept of sustainable development, which assumes reduction of waste and waste and pollution production and choice of goods and services that meet the most ethical, social and environmental criteria. Connecting various entities in the prosumption system, including households, communities and local communities, business communities, local governments, national governments and international structures, may prove to be a significant facilitation in the implementation of the concept of sustainable development.

When discussing the problem of prosumption it should be noted that the phenomenon would not exist if it were not for consumer characteristics and behaviors that trigger and stimulate the discussed processes. Mróz [25] defines consumers of the 21st century as consumers seeking (trysumer), who verify the market offer, are distrustful towards producers and sellers, that is why they make decisions based on their own experiences. Tapscott [31] points out that prosumers are mainly a generation of networks that treat the world not as a place of consumption, but as a place of creation. Therefore, they are not passive consumers, but the creators who are characterized by the need for freedom and freedom of choice, the need to adjust things to their needs and their individualization, the willingness to make joint decisions, the need for dialogue, entertainment and fun, and fast pace and innovation. Prosumers participate with enterprises or other consumers in the process of marketing creation and contribute to the generation of material (products) and intangible effects (image, brand), beneficial to all participants of the relationship created thanks to this cooperation [3].

Along with technological development, there is a change in prosumer behavior—from prosumers whose activity is controlled by companies—giving them individual elements so that they can make a product according to their preferences or modify it—to a prosumer whose activity eludes from the control of companies, consumers on the basis of products create new solutions according to their own ideas and only later companies find out what consumers are doing with their products. Taking into account the different level of prosumer activity one can distinguish among them those whose activity is limited to assessing and evaluating products, those who actively respond to shares organized by producers and participate in activities (aimed at making their offer more attractive, creating a product brand), and also those who for reasons of economy or practicality, design and produce final products, and prosumers of innovators who try to create the offer of companies themselves, design their own products and present their concepts to specific companies [29].

As in many cases the role of the enterprise begins to be limited only to the production and distribution phase of the product, as the other functions are taken over by an intelligent consumer, therefore the literature emphasizes that prosumption creates a real chance to create innovative ideas and achieve higher profits thanks to cost reduction research and development, or building a community of still new users [37]. The argument in favor of prosumption is also the fact that the goods and services that arise in this way are good, especially those that involve many customers-producers at the same time. In many cases, giving the opportunity to make certain decisions to the crowd gives better results than decisions taken by specialists in a given field.

Pointing to the benefits of prosumption, the attention should be paid to the possibilities of using the potential of pros which are managers, engineers or entrepreneurs. Ideas and solutions generated by them often allow to streamline the processes taking place in companies that they know and are responsible for, including supply chain management, optimization of logistics processes, automation of production, and improvement of security standards. The mentioned decision makers also play a key role in initiating activities related to the use of material- and energy-saving solutions, ecological packaging and intelligent packaging.

4 The Role of Prosumers in the Context of Implementing Intelligent Products and Solutions

In business practice many companies use consumer knowledge when creating new products. They are encouraged to share their knowledge and new ideas. Sometimes their knowledge is used to solve specific problems faced by producers. When analyzing the mutual relations between producers and consumers, one can point to the evolution of the approach to consumers as partners in the process of production of consumer goods. Producers through the development of consumer orientation, individualization of products, creation of consumer activity and treating the consumer as a creative resource gain greater consumer satisfaction and loyalty [10].

Active prosumers become the driving force of companies' innovativeness, participating both in the process of continuous search in accordance with the principles of the knowledge-based economy and the development of the consumer society [30]. By analyzing the emergence of modern technologies and the process of their improvement it can be seen that users are the first to usually innovate. Pointing to new solutions they strive to cooperate with producers not only in formal contacts, but also in informal ones, and in addition direct ones. The assumption of a situation where users innovate, and these are not subject to further diffusion, would mean wastage from the point of view of social welfare. Therefore, as part of the new marketing concept, i.e. marketing 4.0, enterprises are obliged to create a cooperation platform, which is necessary to ensure interaction between the traditional seller and the buyer [26]. Only in this way will the prosumer be able to enter into cooperation with the company. The need to create cooperation platforms is indicated by [5], who in prosumption sees sources of innovative solutions, e.g. product, technical and technological, marketing or process ones. In order to confirm the role of prosumers and the aforementioned information exchange system in the process of creating, an attempt was made to develop a model for creating intelligent products using the experience of consumers and producers' resources (Fig. 1).

The smart supply network is supported by ubiquitous data collection networks. It also has system integration not only with trading partners, but also with manufactoring and inventory. Intelligent solutions in supply chain means that the supply network allows to make better decisions by advanced information systems (integrated with intelligent products) and analytical tools.

The proposed model of creating and implementing intelligent products into the market combines the sphere of consumption and production using the information exchange system. The starting point is supposed to be prosumers as a source of inspiring ideas and opinions regarding not the current offer but rather the creation of new solutions. As mentioned earlier, prosumers, unlike ordinary consumers, are characterized by a specific set of features. We are talking not only about formal knowledge, experience resulting from the use of products and practical observations, but in particular about the commitment to solving perceived and anticipated problems, and the desire to creatively communicate and integrate into the creative process of a new product. Features that may also be perceived by prosumers may be

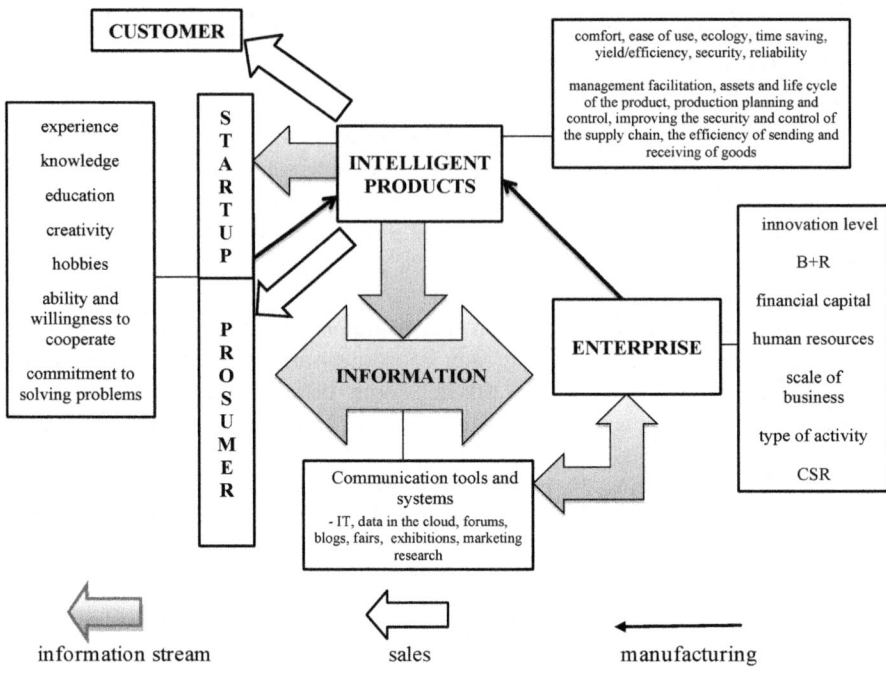

Fig. 1 The prosumer model of creating and implementing intelligent products scheme within the Smart Supply Network

sometimes hobby interest, making the user a specialist with broad knowledge and a good knowledge of the realities and modern technical and technological possibilities. Of particular importance in this model is the desire to cooperate with producers to create a new reality. It is assumed that the prosumer can provide specific knowledge in the area of intelligent products rated as extremely innovative. The model also captures the company not only as an entity that delivers products to the market, but in a broader context looking for ways to achieve a distinctive competitive position. The implementation of intelligent products can serve this purpose. Enterprises with a different scale of activity, having specific financial capital or human capital (bigger ones probably have greater opportunities) may have adequate production potential, but this is not enough if the companies do not have a good idea for the product. Although R&D departments monitor market trends and undertake work on new intelligent products, depending on the industry, the level of innovation in the offer varies considerably. These activities may in turn be insufficient in the effective and efficient creation and implementation of intelligent products. Today some companies use the involvement of prosumers in generating ideas and passing on information about ideas, expectations, potential barriers, and the company receives free expertise. In addition, it is to some extent a guarantee of market acceptance of the intelligent product by the market, because it derives from it. Therefore, in the presented key model there is information, passed from the generator generating intelligent prosumer solutions to those interested in intelligent company products. The materialized concepts of prosumers reach the market and their recipients are both prosumers and "regular" consumers. In this respect, one can point to the social dimension of the involvement of prosumers in social innovation and the dissemination of intelligent technologies. Regular customers will accept smart products as long as they see benefits and usability in the form of comfort, ease of use, a higher level of environmental performance, time savings, higher efficiency and effectiveness, security of use and reliability. If the recipients of intelligent solutions are enterprises, links in the supply chain, the expected benefits are all solutions facilitating and improving the management of the production process, management of raw materials and finished products deliveries, a higher level of control and flow safety or rationalization enabling cost reduction. If the partner/customers have special expectations then smart products may even be necessary for the continuation of cooperation in business. This may refer to the need to deliver smart packaging informing about the current state of the product (e.g. food) or intelligently organic packaging, biodegradable after using the product itself—as new standards required from packaging suppliers. Prosumers, on the other hand, gain additional satisfaction from active participation in creating reality, appreciating the meaning of their activities and cooperation with manufacturers, they continue to engage and provide further ideas. In the model, we also drew attention to the fact that it is necessary to develop tools and systems of communication with clients, streamlining the transfer of information between prosumers and the company.

Numerous solutions implemented in this area are already visible today by enterprises which, due to their obvious benefits, are interested in developing these ties. Their implementation may be more or less formal (fora, blogs, marketing research), personal or anonymous (fairs, exhibitions, solutions based on Internet). The specific entity included in the presented model are startups. They are not strictly entrepreneurs, but rather entities that are in the phase of organization, which are based on an innovative idea. Startups show a number of features identical to prosumers, because in fact their role in the concept of creating a product concept is similar. These startups are different from ordinary prosumers, that they are trying to implement their own ideas, commercialize them and not just forward suggestions or information to implementing companies. Startups, however, often do not show durability and decline, which can be attributed to the lack of sufficient resources characteristic of classical enterprises and the weaknesses of products or market risks that the founders have not noticed. It does not change the fact that the characteristic associated with startups is the innovation of the concept of products and services, with frequent implementation of automated, mobile or artificial intelligence in products.

5 Negative Aspects of Prosumption Development and Creation of Intelligent Products

It follows from the considerations that the phenomenon of prosumption has a significant impact on the creation of new relations between producers and consumers, which results in new solutions and products, including intelligent products. It should be noted, however, that despite many undisputed benefits resulting from the development of prosperity and intelligent products, this process is accompanied by negative effects, which lie in the prosumption itself as well as in smart products resulting from prosumer activities.

Some researchers do not share optimistic views regarding the legitimacy of prosumers and draws attention to the dangers associated with this trend. This view is emphasized by Kopeć [18] or Bardhi and Eckhardt [2] perceiving the phenomenon of prosumption as a new form of exploitation. The author points to three areas in the producer-consumer relationship that lead to the use of consumers for purely economic purposes. The first of these is the transfer of some of the tasks carried out so far by producers to consumers. An example may be the process initiated by introducing various forms of self-service, popularizing DIY solutions initiated in the furniture sector (IKEA) or other forms of activity on which the modern online economy is based—e.g. booking travel, ordering books, booking tickets to the theater or making bank transfers. This problem is noticed by Fuchs [13] who accuses A. Toffler that he overlooked some prosumption defects, which are outsourcing of work, which was previously performed by employees and handed over to customers without any payment for it.

The second area where consumers are exploited is the creation of value products and services based on content based on consumer creativity (e.g. consumer opinions, video, photos, conversations). By conducting discussions on the web or publishing comments about the purchase of a given product or a service, prosumers indicate solutions to many technical problems, which voluntarily contribute to a certain value, which may become the basis for profit for companies. In addition, positive opinions about the product of a given company contribute to its propagation and thus to increased sales of a given product. In this sense, the exchange of information on the web becomes, as a result, unpaid, additional work, and so-called co-creation can be a form of exploitation because it is not seen as part of the consumption process, but rather as part of the production process, offering consumers little more than the illusion of freedom and power [40].

The third controversial area of prosumption concerns the problem of collecting and using information about the user, their behavior and the effects of activities on the network that are collected by a third party [12]. Active prosumers provide themselves as users of social networking sites like Twitter, Facebook or MySpace, providing business entities, owners of these portals with rich information that enable the creation of their detailed profiles based on observation and analysis of customs and activities undertaken on the Internet, search history, completed purchases, membership of various member groups [10]. The effects of the user's activity in the network, including entries, photos or personal data, are archived and can be found by other entities (marketing companies, government agencies). The collected data about consumers becomes a commodity then [19], and prosumption itself is not a form of sharing power, but rather a new means of exploitation and control under the guise of freedom given to the Internet user.

Another group of threats are hazards associated with intelligent products themselves and their impact on individual consumers and the whole society. As it turns out consumers do not appreciate intelligent products for their intelligence itself, but because of the relative advantage and compatibility that they deliver. Also, consumers have a lower appreciation for intelligent products when they perceive these products as more complex [29]. Besides many advantages involved with intelligent products they also have some disadvantages. Intelligent products may contain hidden functionality which customers do not want especially if they are unaware of it. Also intelligent products may be perceived as complex, as a consequence of its innovations stage and automatic functions. The unique capabilities of intelligent products may encounter resistance from customers in the form of perceived risk resulting of using and doubts about their real intelligence [28].

The issues of complexity of operation as well as the built-in and hidden functions of intelligent products raise the attention to doubts raised by customers as to the direct security of their use and the security of data that these products collect and transmit. Another unfavorable consequences of implementing intelligent products from the perspective are related to their price. As highly innovative products, they are especially expensive in the initial phase of the life cycle.

Finally, the impact of intelligent products on people remains a delicate issue. The idea of freeing man from engaging the thought process and physical effort is at the heart of intelligent products. Nevertheless, one should reflect on the long-term effects of using intelligent products. This is not about the extreme opinion that the devices were to take control of a human but about the possible loss of certain abilities and skills of the user after a certain time (as a result of the intelligent products), which can be a serious problem in coping with basic problems in emergency or crisis situations.

Without prejudging the final balance of profits and losses, one should also pay attention to the fact that the implementation of new systems, devices, sensors and software into intelligent products should also be considered from the ecological perspective, i.e. energy consumption and the use of resources and raw materials used for their production. Some innovative products—which provide higher performance or greater functionality and are identified with lower energy consumption or lower gas emissions—are accused of high environmental burden in other areas (total emissions of electric cars in the production, exploitation and disposal phases). The problems and doubts mentioned do not overshadow all benefits and will not change the direction of products and services development towards advanced information technologies and intelligent solutions. However, they should be considered as areas that should be considered by the producers in the strategies of introducing intelligent products into the market, not avoiding them but referring to them as part of marketing communication, or if necessary to use them in the process of designing and manufacturing of intelligent products.

6 Conclusions

Smart Supply Network is becoming more effective and efficient by using of digital technologies. SSN needs to create an integrated information system with customers, suppliers and partners and exchanging data with them, thereby increasing the transparency of the entire supply network. Information feeds are invaluable for enterprises, they condition their development and risk reduction. This is evidenced by the growing importance of marketing research, the improvement of data collection and processing techniques, and the search for access to information sources. Modern enterprises are increasingly engaging prosumers to gain knowledge about new solutions, including in particular those where intelligent solutions are implemented. For many companies prosumers become additional resources with experience, the ability to think creatively, which not only complements the efforts of R&D departments of companies but because of the double role of an expert and customer they are a natural verifier of proposed solutions and a source of information about the acceptance of intelligent products. As the economic reality proves, the idea itself is often too little, which is manifested by the low level of market survival of startups, often possessing a set of features similar to prosumers, and the need for appropriate financial and human resources, technological support—which is the domain of producers. Both prosumption and intelligent products are intended to serve consumers and facilitate their operation, provide time savings, provide new values, streamline production and optimize flows. However, it should be emphasized that these activities are accompanied by negative consequences and the same features of intelligent products may raise doubts of potential customers. They concern the degree of complexity and embedded and hidden functions, process control carried out by intelligent products, safety and care for the superiority of final results over those provided by conventional products. The issue of data security, which intelligent products collect and transfer, is also growing. Taking into account the fact that prosumption is a phenomenon that is strongly evolving and showing an upward trend, it is necessary to undertake actions related to limiting the negative impact of this process on prosumers and the whole society. One cannot allow the development of prosumption and intelligent products to take place at the expense of exploitation, surveillance and control of those who are the main source of inspiration and solutions implemented by companies. Therefore, it becomes necessary to develop mechanisms and standards that enable the potential of prosumers to be used, while maintaining all rights related to the protection of intellectual property and respect for freedom. An important problem that requires further research and analysis is also the issue of better use of the potential of startups, led by prosumers, interested in independent implementation of ideas, as well as the improvement of systems and tools for communication of companies with prosumers.

References

1. Barańska-Fischer, M. (2016). Innowacyjne produkty systemowe jako efekt implementacji inteligentnych specjalizacji regionu (Innovative system products as a result of the implementation of smart specializations of the region), Studia Ekonomiczne. *Zeszyty Naukowe Uniwersytetu Ekonomicznego w Katowicach, 262,* 116.
2. Bardhi, F., & Eckhardt, G. M. (2017). Liquid Consumption. *Journal of Consumer Research, 44,* 591.
3. Baruk, A. (2017). Prosumpcja jako wielowymiarowe zachowanie rynkowe (Prosumption as multidimensional market behavior), PWE Warszawa, p. 87.
4. Battistella, C., & Nonino, F. (2012). Open innovation web-based platforms: The impact of different forms of motivation on collaboration. *Innovation: Management Policy & Practice, 14,* 557–575.
5. Bembenek, B. (2016). Strategiczny wymiar prosumpcji w klastrach wiedzy (Strategic dimension of prosumption in knowledge clusters). Handel wewnętrzny 1, 187–201.
6. Bendkowski, J. (2016). Tworzenie wiedzy i uczenie się w warunkach nowej organizacji pracy w gospodarce wirtualnej (Creating knowledge and learning in a new work organization in a virtual economy), Zeszyty Naukowe Politechniki Śląskiej, seria Organizacja i Zarządzanie, no. 95, s. 24.
7. Berthon, P. R. (2012). Marketing meets Web 2.0, social media and creative consumers: Implications for international marketing strategy. *Business Horizons, 55*(3), 261–271.
8. Boulaalam, A., Nfaoui, E. H., & Beqqali, O. E. (2013). Intelligent product: Mobile agent architecture integrating the end of life cycle (EOL) for minimizing the lunch phase PLM., The International Journal of Soft Computing and Software Engineering [JSCSE] 3(3 Special Issue), 109.
9. Bylok, F. (2014). Prosumpcja na rynku energii w perspektywie teoretycznej, [in] *Energetyka prosumencka—pierwsza próba konsolidacji* (Prosumption on the energy market in the theoretical perspective, [in] Prosumer energy—the first attempt to consolidate), red. J. Popczyk, R. Kucęba, K. Dębowski, W. Jędrzejczyk, Sekcja Wydawnictw Wydziału Zarządzania Politechniki Częstochowskiej, pp. 90–93.
10. Bylok, F. (2015). Prosumpcja jako forma innowacji na współczesnym rynku (Prosumption as a form of innovation in the modern market). *Marketing i Rynek, 2,* 55.
11. Claffey, E., & Brady, M. (2014). A model of consumer engagement in a virtual customer environment. *Journal of Customer Behaviour, 13*(4), 325–346.
12. Cohen, N. S. (2008). The valorization of surveillance: Towards political economy of Facebook. *Democratic Communiqué, 22,* 5–22.
13. Fuchs, C. (2011). Web 2.0, Prosumption and surveillance. *Surveillance & Society, 8*(3), 291–297.
14. Guttman, U., Papst, J., Merlo, R., Kane, D., Bieser, G., & Grob, O. (2017). Industry 4.0: What's Next. In *An SAP Point of View, SAP, White Paper, Industry 4.0,* p. 5, September 2017.
15. Ives, B., & Vitale, M. R. (1988). After the sale: Leveraging maintenance with information technology. *Management Information Systems, Quarterly, 12*(1), 7–21.
16. Kärkkäinen, M., Holmström, J., Främling, K., & Artto, K. (2003). Intelligent products—A step towards a more effective project delivery chain. *Computers in Industry, 50*(2), 141–151.
17. Katana, K. (2016). Innowacje społeczne w kontekście rozwoju inteligentnych organizacji (Social innovations in the context of the development of intelligent organizations). *Zeszyty Naukowe Politechniki Śląskiej, seria Organizacja i Zarządzanie, j., 92,* 152.
18. Kopeć, K. (2014). Empowerment czy wyzysk? O niejednoznacznej naturze prosumpcji w sieci (Empowerment or exploitation? About the ambiguous nature of prosumption on the web), Kultura i Polityka, Zeszyty Naukowe Wyższej Szkoły Europejskiej im. ks. *Józefa Tischnera w Krakowie, 16,* 158–170.

19. Kuehn, K. M. (2011). Prosumer-citizenship and the local: A critical case study of consumer reviewing on Yelp.com, The Pennsylvania State University, The Graduate School, College of Communications, p. 44. https://etda.libraries.psu.edu/paper/12069/7819.
20. Maleszka, M., & Nguyen, N. T. (2015). Integration computing and collective intelligence. *Expert Systems with Applications, 42*(1), 332–340.
21. McFarlane, D. (2012). Product Intelligence: Theory and practice. *IFAC Proceedings Volumes, 45*(6), 11–12.
22. McFarlane, D., Sarma, S., Chirn, J. L., Wong, C. Y., & Ashton, K. (2003). Auto id systems and intelligent manufacturing control. *Engineering Applications of Articial Intelligence, 16* (4), 365–376.
23. Meyer G. G., Främling, K., & Holmström, J. (2009). Intelligent products: A survey. Computers in Industry 60(3), 6, 137–148.
24. Mitręga, M. (2013). Czy prosumpcja w dobie kryzysu to zjawisko jednowymiarowe? Eksploracja wśród użytkowników portali społecznościowych (Is prosperity in the time of crisis a one-dimensional phenomenon? Exploration among users of social networks), Problemy Zarządzania 11(1, 40), 44.
25. Mróz B. (2010). *Nowe trendy konsumenckie—szansa czy wyzwanie dla marketingu* (New consumer trends—an opportunity or a challenge for marketing) [in:] Marketing w realiach współczesnego rynku. Implikacje otoczenia rynkowego, red. Sz. Figiel, PWE, Warszawa, p. 64.
26. Nowacki, P. (2014). Marketing 4.0—nowa koncepcja w obliczu przemian współczesnego konsumenta (Marketing 4.0—a new concept in the face of modern consumer changes). *Marketing i Rynek, 6,* 14–18.
27. Porter, M. E., & Heppelmann, J. E. (2015). How intelligent, connected products are transforming companies. https://hbr.org/2015/10
28. Rijsdijk S. A., & Hultink E. J. (2013). Developing intelligent products. In K. B. Kahn (Ed), *The PDMA handbook of new product development* (3rd Ed., p. 301). London: Wiley.
29. Rijsdijk, S. A., Hultink, E. J., & Diamantopoulos, A. (2007). Product intelligence: Its conceptualization, measurement and impact on consumer satisfaction. *Journal of the Academy of Marketing, 35,* 343–353.
30. Szul, E. (2013). Prosumpcja jako aktywność współczesnych konsumentów—uwarunkowania i przejawy (Prosumption as activity of contemporary consumers—Conditions and manifestations). *Nierówności społeczne a wzrost gospodarczy Nr., 31,* 347–356.
31. Tapscott, D. (2010). *Cyfrowa dorosłość. Jak pokolenie sieci zmienia nasz świat* (Digital adulthood. As the generation of the network changes our world). Warszawa: Wydawnictwa Akademickie i Profesjonalne.
32. Tapscott, D., & Williams, A. D. (2008). *Wikinomia. O globalnej współpracy, która wszystko zmienia* (Wikinomia. About global cooperation that changes everything) (pp. 215–216). Warszawa: Wydawnictwa Akademickie i Profesjonalne.
33. Toffler, A. (1997). *Trzecia fala (Third wave)* (pp. 43–45, 409–412), Warszawa: Państwowy Instytut Wydawniczy.
34. Vallverdú, J. (2014). Handbook of research on synthesizing human emotion in intelligent systems and robotics, IGI Global (p. 177).
35. Wierzbiński, B., & Surmacz, T. (2012). Advantages of collaborative approach in customer service management. *Research in Logistics & Production, 2*(1), 115–126.
36. Wang, C., Chen, L., Mengyan, L., & Zhao, L. (2013). Exploring the norms for the UX design of intelligent products: A case study, design management symposium (TIDMS). In *2013 IEEE Tsinghua International* (p. 164).
37. Wolny, W. (2012). *Prosumpcja—konsumencka kreatywność w gospodarce elektronicznej* (Prosumption—consumer creativity in the electronic economy), Zeszyty Naukowe Uniwersytetu Szczecińskiego, no. 703, Ekonomiczne Problemy Usług, no. 88, p. 124.

38. Wolny, W. (2013). *Formy prosumpcji w systemach informatycznych, w: Innowacje w zarządzaniu i inżynierii produkcji* (Forms of prosumption in information systems. In: Innovations in management and production engineering), red. R. Knosala, Oficyna Wydawnicza Polskiego) Towarzystwa Zarządzania Produkcją, Opole, p. 950.
39. Xie, C., Bagozzi, R. P., & Troye, S. V. (2008). Trying to prosume: Toward a theory of consumers as co-creators of value. *Journal of the Academy of Marketing Science, 36*(1), 110.
40. Zwick, D., Bonsu, S. K., & Darmody, A. (2008). Putting consumers to work: Co-creation and new marketing govern-mentality. *Journal of Consumer Culture, 8*(2), 177.

Printed by Printforce, the Netherlands